THE SCIENTIST & THE SOCIOPATH

BY JOSEPH D'AGNESE

THE SCIENTIST & THE SOCIOPATH presents real-life stories about scientists and laymen tackling theories and discoveries that will change our world.

Culled from the author's reporting for magazines such as *Discover, Wired* and *Seed*, these tales are presented together for the very first time. This ebook includes 16 nonfiction stories and a selection of the author's science book reviews.

GET THE NEWSLETTER

Signing up for the newsletter entitles you to receive a free Starter Library of the author's books, not to mention advance news on upcoming books and specials. See the back of this book for details or visit the website.

JOSEPHDAGNESE.COM

ALSO BY JOSEPH D'AGNESE

LEARN MORE AT JOSEPHDAGNESE.COM

FICTION

Jersey Heat

The Mesmerist

Ear of God

The Marshal of the Borgo

Murder on Book Row

Sorceress Kringle

The Icemaster of New-York

Arm of Darkness (short stories)

Daggyland collections (short stories)

NONFICTION

Signing Their Lives Away

Signing Their Rights Away

Stuff Every American Should Know

The Money Book for Freelancers

Speaking My Sleuth collections (essays)

The Indiana Jones Handbook

FOR CHILDREN

Blockhead: The Life of Fibonacci

American History Comic Books

THE SCIENTIST & THE SOCIOPATH

True Science Stories

JOSEPH D'AGNESE

NutGraf

The Scientist and the Sociopath

Cover design by Jeroen ten Berge

josephdagnese.com

Dedicated to the memory of scientists Ralph Alpher, Alan Thorne, and Paul MacCready. I am touched still by their kindness and generosity.

CONTENTS

Introduction vii

The Scientist & the Sociopath 1
Last of the Big Bang Man 14
Where Chimps Go to Die 28
The Medical Detective 40
Yes, Snakes Can Fly! 53
The Organ Makers 59
Organs in the Shop 73
The Lady in the Suitcase 77
The Scientist Who Threw the Light Switch 87
The Weather Inside You 100
Patriots Aboard the Big Blue Bus 108
The Town Changed By Wind 113
Mars Bound 124
Do More With Less 129
Know Thyself 135
Worshipping at the Church of the Opaque 142
Book and Media Reviews 162

A Request from the Author 176
Acknowledgments 177
About the Author 178

INTRODUCTION

Most of the stories in this collection are profiles about real-life scientists or physicians, and the work they do. Two stories are about green energy. One's about space travel.

All of the stories are true.

Some of the people in these stories are not scientists at all. They are laymen grappling with some aspect of science in their lives. Along the way, there are smashed ancient skulls, dead chimps in the back of pickup trucks, flying snakes, lordly windmills, haunted warriors, and beautiful, geeky kids building us a new world, one Lego at a time.

One note seems in order. This is a work of nonfiction, and thus a snapshot of a moment in time. Since these articles first appeared in print, individuals have revised their scientific theories, moved to different institutions, have retired or even died. Articles appear largely as they did when originally published. Dates and notes accompanying each article provide some context and updated information.

Joseph D'Agnese
March 2011

THE SCIENTIST & THE SOCIOPATH

2004

A mysterious manuscript written in a strange, indecipherable alphabet. What could be more tantalizing? But for one scientist, the bizarre book is just a gateway to cracking bigger mysteries.

Two years ago, a Scotsman named Gordon Rugg slipped back in time. Night after night he spread his papers on the kitchen table once his children had gone to bed. Working on faux parchment with a steel-nibbed calligraphic pen, he scribbled a strange, unidentifiable, vaguely medieval script. If you converted some of those letters into the alphabet we use, some of the words would read: *"qopchedy qokedydy qokoloky qokeedy qokedy shedy."* As he wrote, Rugg struggled to get inside the mind of the person who had first scrawled this incomprehensible text some hundreds of years ago.

By day, Rugg, a forty-eight-year-old psychologist, teaches in the computer science department of Keele University, near Manchester, England. By night, as an intellectual exercise, he has been researching one of the world's great oddities: the Voynich manuscript, a hand-lettered book written in an unknown code that has frustrated cryptographers since its discovery in an Italian villa in 1912. How impregnable is the Voynich? During World

War II, code breakers who blew away Nazi ciphers grappled with the manuscript in their spare time and came up empty. Since then, decoding the book's contents has become an obsession for geeks and puzzle nuts everywhere.

Then came Rugg. In three months, he cooked up the most persuasive explanation yet for the 234-page text: Sorry, folks, there is no code—it's a hoax! Some Voynichologists were impressed with his reasoning and proofs, even if they were a little chagrined. Others, who disagreed with his findings, were appalled that his work got so much press, nabbing coverage in top journals such as *Nature* or news outlets like the BBC. "The Voynich is such a challenge," says Rugg, "such a social activity. But then along comes someone who says 'Oh, it's just a lot of meaningless gibberish.' It's as if we're all surfers, and the sea has dried up."

When the news of Rugg's breakthrough was published in January 2004, everyone missed the bigger story. Rugg's finding came not because he was smarter, but because he focused on what everyone else had missed. This came naturally to Rugg: He has made a career out of studying how experts acquire knowledge yet screw up nevertheless. In 1996, he and his colleagues developed a rigorous method for peering over the shoulders of experts—doctors, software engineers, pilots, physicists—watching how they work and think, testing their logic, and uncovering ways to help them solve problems.

Rugg calls this method the verifier approach, and the Voynich was its first major test. If Rugg gets his way, verifiers will revolutionize the scientific method and help solve other seemingly unsolvable mysteries, such as the origins of the universe or the cause of Alzheimer's disease.

~

RUGG WAS HARDLY THE FIRST TO DREAM OF CRACKING the Voynich. Ever since the manuscript resurfaced—bookseller Wilfrid Voynich bought it from Italian Jesuits ninety-two years ago—a stream of formidable scholars have pored over it. Some make pilgrimages to Yale's Beinecke Rare Book and Manuscript Library, where the volume resides. Others download JPEGs of the pages, which are available free on the web.

Rugg saw something different and special about the manuscript: It would make a perfect beta test for the verifier approach. As he read about the Voynich and began applying his method—amassing knowledge about a problem and assessing the kinds of expertise applied so far are steps one and two—he saw that no one had seriously explored the idea that the book was a grand hoax. As Philip Neal, one of the world's leading Voynichologists, says, "It has been argued—I used to argue myself—that the phonetic structure was beyond the powers of a sixteenth-century forger to create, so that the text must be a real language or an unknown type of cipher."

Since none of the experts thought a hoax was plausible, no one had looked very hard for a hoax solution. To compound the problem, many Voynichologists were specialists: linguists, cryptographers, mathematicians, medievalists, and literary scholars. But the ideal Voynich expert—a code-breaking, medieval-savvy hoaxologist—probably didn't exist. And the resulting gap had allowed a major problem to go unsolved for the better part of a century.

This "expertise gap" is rife in academia, but few recognize it, let alone know how to correct for it. It starts with the best of intentions. Institutions want top-notch people, so they offer incentives to attract and groom experts. Young grad students learn early that if they want to carve out a niche, they must confine their interests to a narrow field. It's not enough to work in spinal cord regeneration; it must be stem cell-based solutions to the problem. That's great if a researcher just happens to stumble on a perfect stem cell cure. But as specialists get further from their core expertise, the possible solutions—what's been tried, what hasn't, what was never properly examined, what ought to be tried again—get even more elusive.

With the verifier approach, Rugg begins by asking experts to draw a mental map of their field. From there, he stitches together many maps to form an atlas of the universe of knowledge on the subject. "You look for an area of overlap that doesn't contain much detail," he says. "If it turns out there's an adjoining area which everyone thinks is someone else's territory, then that's a potential gap."

Rugg asked himself: If I were living in the sixteenth century and wanted to make a book that looked mysterious but was really gibberish, how could I do it cheaply and easily? He deliberately searched for low-tech tools capable of generating text that seemed random. In his reading, he came across an encoding device called the Cardan Grille, first described in 1550 by Girolamo Cardano. The methodology is fairly simple. You fill in a large grid with randomly generated word fragments, called prefixes, midfixes, and suffixes. The word fragments can be anything, nonsense syllables written in a fake alphabet, signifying nothing. Next, using heavy card stock, you cut a three-slot grille that exposes these word fragments. Then you work your way through the grid, sliding the grille over three cells to form a word. Each time you do this, you copy the new "word" you've just formed into the manuscript page. You continue moving the grille around the grid. Just to shake things up, you occasionally cut a new grille out of card stock, and repeat the whole process.

Using such an encoder, Rugg figures it would have taken a smart fraudster an hour or two to write an entire page. A Voynich-size book might have taken about three or four months to create with illustrations. The time and effort would have definitely been worth it: In the Elizabethan era, Rudolph II, the Holy Roman emperor, became fascinated with the beautifully wrought manuscript (he believed it was the work of the thirteenth-century philosopher Roger Bacon) and paid six hundred gold ducats for it—about $30,000 today.

The text's author was long-dead. Rugg couldn't watch him work, but he could get inside his head by trying to replicate his pen-and-ink technique. The precision paid off. Once, when the ink blotched, Rugg swore aloud and thought about discarding the grid he had so laboriously written on a big sheet of paper. Then it struck him that the author must have experienced the same thing. What was the best solution? Toss it out? No. Paper was expensive. A new grid? No, too much work. Better to leave the blotch and work around it. That helped him realize that some of the cells in the original grids must have been left blank—the resulting missing syllables yielded a variety of word lengths, giving the faked language even more verisimilitude.

Rugg thinks the hoaxster could have been a man named Edward Kelley, a mercurial, shape-shifting villain who walked into history on March 8, 1582. He vanished from history fourteen years later, as abruptly as he had entered. No one knows what happened to him, and it's unlikely anyone ever will. He probably wanted it that way.

Rugg has spent a lot of time thinking about Kelley, getting inside his head, trying to make the pen in his hand move as Kelley's may have, and the sociopathic con man has proved to be a remarkably elusive quarry.

It's easy to imagine Kelley marching out of the mist on some night when the weather is bad in the Shropshire countryside, where Rugg lives. The apparition is stooped, partially lame in one leg, his head obscured by a hood, probably used to conceal a deformity of one ear. As he approaches, the hood drops and you can see the ear for what it probably was, a horrible disfigurement inflicted upon him as punishment for the crime of counterfeiting. The man is unshaven, filthy, dressed in rags. He snarls, and disappears into the mist again, but not before his aspect and clothing transform entirely. No longer is he dressed like a man on the fringes of society. Now he wears the robes of a magus, a wise man, and then, just before he vanishes, the finery of a wealthy baron.

This is the way Kelley appears to historians, too. They can't quite draw a bead on him because the records depict him in various roles—a defrocked monk, a necromancer, a soldier of fortune. Was he all of these things or none of them? Or did he simply pull on different disguises according to the con he was working at the time?

In 1582, Queen Elizabeth I's court scholar, Dr. John Dee, heard a knock at his door. An advisor to Queen Elizabeth, a scholar with an international reputation, Dee was a hard-headed administrator who had been summoned by the queen to organize and restructure the national dockyards at a time when the navy was the country's main hope in the threatened war with Spain, the superpower of the time. Dee's house in the Mortlake region of London was an imposing building, befitting a major figure of the Elizabethan world. That night in 1582, Dee opened the door to find a man who claimed to be a "skryer," a clairvoyant who supposedly can divine truth from crystal balls. The man was Kelley, who hoped to insinuate himself into Dee's household, and gain access to court life and the international intelligentsia with whom Dee associated. The usual depiction of the scene is of a demonic Mephistopheles meeting a credulous Faust, but the reality is far more complex, and in many ways stranger than its fictional counterpart.

At this stage in history, the line between science and religion was porous. Men of science like Dee were intrigued by both magic and spirituality. Dee's interests were right on the cusp of the real and the bizarre, the rational and the hocus-bogus. Yet he was also a deeply religious man, and one of Europe's leading intellectuals. He reasoned that if God had communicated with mortals via angels in the Bible, it stood to reason that angels could still transmit God's will to mankind.

From the moment he entered Dee's home, Kelley began working a masterful long con that poor Dee never saw coming. Kelley convinced Dee that he could speak to angels. During their séances, Kelley communicated in a bizarre tongue called Enochian. To this day, scholars admire the richness of this obviously fabricated language. It was impenetrable enough to fool Dee, whose knowledge of cryptography was vast. Next Kelley produced a treasure map, which he claimed was written by angels. Dee managed to decode the language in which the map was written, but the instructions were too vague to allow the men to actually find the holy treasure. This episode, says Rugg, could have been a devious test of Dee's code-breaking skill. The angels, through Kelley, next instructed the men to have sex with each other's wives. Dee's beautiful young wife balked, but Dee managed to convince her to carry out the angelic command, damaging their marriage in the process.

Bit by bit, Kelley pushed the envelope, upping the stakes into previously uncharted territory. At the time, Dee had been hiring clairvoyants left and right, trying to divine what these mysterious psychics knew about the universe. He'd seen so many charlatans that he'd become cynical about them, and not easily fooled. The fact that Kelley managed to fool him says a lot more about Kelley's skill than about Dee's credulity.

At some point the two men journeyed to Bohemia to meet with the Holy Roman Emperor. Rugg thinks Kelley may have crafted the mysterious manuscript specifically to bilk the Holy Roman Emperor Rudolph II, a credulous patron of arts who was known to buy just about anything under the sun. (Rudolph once bought a relic described as "the skull of John the Baptist, as a young boy.") Obsessed with the occult, Rudolph spent a lifetime in search of the Philosopher's Stone, a magical object believed to turn worthless metal into gold.

Historians have long known that Rudolph paid a fortune for the Voynich manuscript, and they have even speculated as to who the seller was. But Rugg's hoax theory, if true, fills in a lot of the blanks. One possibility is that Kelley forged the manuscript, and had a fellow conspirator offer the book to Dee, and mention that they could sell it to Rudolph, but only if it could be verified by an expert.

"Kelley would have seen Dee in action," says Rugg. "He knew Dee could decode anything. He had a healthy respect for the man's ability. Kelley was deviously smart. He tricked a lot of smart people in his time. I have a lot of grudging respect for the man."

Dee may have volunteered to vet the text, using his considerable code-breaking prowess to decipher the material, only to fail. That experience may have convinced him that the book contained incalculable secrets. The trio may have then presented the book to the Emperor. If Dee vouched for it, all three men would have then pocketed handsome commissions when the sale was completed. Dee would have been grateful to Kelley for his go-between share of the proceeds, not realizing that the apprentice was starting to overshadow the master.

Is this theory provable? Probably not. But it is provocative because it's the only one that fits all the facts that are undisputed. Rudolph paid six hundred ducats for the book. Kelley had a talent for creating superb secret codes and a history of manipulating Dee. Shortly after meeting Rudolph, Dee and Kelley lived in luxury off a large sum of money, the source of which has never been identified. The artwork in the Voynich manuscript exhibit Kelley's brand of sophomoric eroticism. The frolicking nudes, specifically, appear to be the work of a man who can barely keep a lid on his oversexed mind.

Eventually Dee, cut off from a royal stipend upon Queen Elizabeth's death, fell into poverty. Kelley's star rose and he became wealthy. He traveled the continent, continuing a string of manipulations and con jobs, possibly under the protection of Rudolph, who now believed him to be a true alchemist. The emperor named him a baron, but may have eventually tired of waiting for his huge payday. Kelley was twice imprisoned, the last time in a castle in the city of Most, now in the Czech Republic. Kelley shape-shifted again: drugged his guards and escaped through a cell window on a rope made of bedclothes and vanished from history.

Colorful as this theory is, it may be impossible to prove Kelley's involvement. That doesn't trouble Rugg. His chief contribution to the Voynich literature—indeed, the whole point of his academic paper—is demonstrating that a hoax is feasible. That assertion doesn't depend on Kelley in the slightest. If Kelley didn't write the book, says Rugg, then some other hoaxer or con man did.

Rugg published his findings in the journal *Cryptologia*, a peer-reviewed academic journal. The paper was bolstered by Laura Aylward, one of Rugg's students, who used software to replicate his work. She found that the Voynich's well-known statistical anomalies—its unnatural repetition of certain words, a sudden drop-off in the use of previously common syllables—can all be accounted for if one uses structured tables and grilles. Rugg says his answer is the simplest explanation for the manuscript. Still, many Voynich scholars remain unconvinced that the book was created in this way and for this purpose. "I know how they feel," Rugg says. "The rational part of me says it's a hoax; another part says, yes, but what if 10 percent of it is cipher text, a real message mixed in with all the wattle and padding? It's a lovely problem."

The verifier approach has been applied successfully to only one lovely problem so far—the Voynich—but Rugg is already preparing to tackle more. "We have to be careful," he says. "We are not going to barge in on people. We want to approach them with humility and tact, so they'll see us as partners."

Rugg's approach is built on the observation, noted as far back as the 1970s, that experts tend to cut to the chase. In their zeal to get to an answer, they make many little mistakes, some good, some bad. (A recent study of work published in *Nature* and *British Medical Journal*, for example, found that 11 percent of scientific papers had serious statistical errors.) Experts unknowingly fudge logic to square data with their hypotheses. Or they develop blind spots after years of working in isolation. They lose their ability to take a broader view. If all this is true, he says, think of how much big science is based on flawed intuition.

He gestures as if he's sculpting something invisible yet potent. Rugg's face is youthful—sharp nose, prominent forehead, all brain. He's also a paragon of well-roundedness: He studied French and linguistics, taught English in Nepal, dabbled in archaeology, and wrote a paper on the occurrence of left- and right-handedness in Stone Age peoples. Then came psychology.

His work bridges a number of specialties. One of the tools in his "toolkit," as he says, is a field called judgment and decision-making. Psychological studies suggest that experts, defined as someone with ten years in a discipline, don't have any more intelligence or reasoning power than the rest of us. What they have is tons of experience. An old doctor, for instance, has seen so many cases of the mumps that he no longer follows methodical reasoning to arrive at a diagnosis. He instead uses a shortcut called pattern-matching. "Face red and swollen? Mumps. Next!"

Call it what you like—a hunch, an opinion—pattern-matching is iffy. "Sequential reasoning is formal, almost mathematical," Rugg says as we settle down in the campus cafeteria over mochas. "If this, then that. Pattern-matching is fast and efficient. The doctor knows what's wrong with you before you do. That's fine if he's right, but he can be miscued."

Besides pattern-matching incorrectly, experts sometimes misunderstand critical terms. People in similar but specialized fields will find it hard to communicate. A hydrogeologist and a petroleum engineer both took Geology 101, yet their visceral understanding of the word "rock" may be wildly divergent. Luckily, psychologists have developed a battery of methods—called elicitation techniques—to draw out and define what experts know.

As a journeyman researcher in expert reasoning, Rugg ferreted out errors and unseen problems in various industrial and office management cases. It was good work for a human-error psychologist, but he wanted to tackle bigger issues. If experts were making mistakes in doctor's offices and factories, he reasoned, they were making them in labs, too. "My gut feeling is that a lot of research involves pattern-matching," Rugg says. "It guides what is investigated and then the design of the study."

Sometime in 1996, while having lunch with colleague Joanne Hyde, it occurred to Rugg that he could pull together all the tools psychologists use—elicitation techniques, the vast literature on human error, decision-making models, formal logic and reasoning—to create a novel form of problem-solving: a scientific method to verify the methods of science.

The verifier method boils down to seven steps: 1) amass knowledge of a discipline through interviews and reading; 2) determine whether critical expertise has yet to be applied in the field; 3) look for biases and mistakenly held assumptions in the research; 4) analyze jargon to uncover differing definitions of key terms; 5) check for classic mistakes using human-error tools; 6) follow the errors as they ripple through underlying assumptions; 7) suggest new avenues for research that emerge from steps one through six.

Experts want to believe that their domain is unique, requiring specialized tools, approaches, and thinking. Rugg says no, you can use one kit to solve a million problems, in many fields. Many experts would object to such a theory, but computer scientists are intrigued. "In computing, we're interested in understanding knowledge," says Bashar Nuseibeh, a UK researcher whose work was influential in the development of the Unified Modeling Language. "Gordon is asking, Can you look at the commonality between two domains of research and solve problems within them with a single approach? I don't know. It's a theory."

Equipped with his verifier toolkit, Rugg tried finding a wealthy partner to get his new method off the ground. At least one pharmaceutical company turned him down. "It was a Catch-22," Rugg says. "No one would take a chance on us unless we showed them we could do it first. But I couldn't show I could do it until someone let me test it." That, of course, is where the Voynich manuscript came in. Now he's poised to tackle a much bigger question.

PETER CROME IS THE DEPUTY HEAD OF KEELE'S medical school and a professor of geriatrics. When he smiles, his eyes narrow to jovial slits. If I were slowly losing my mind, I think I would want Crome to break it to me. "Alzheimer's is a mystery," he says. "An odd one. Although we know a lot about it, we don't know what causes it. All we can say at this point to someone we think has it is, Come back next year. If you're worse, you have it."

Alzheimer's is a classic candidate for the verifier approach: It's been studied by tens of thousands of scientists, yet still defies basic definition. Dozens of causes are under investigation: Could it be toxins in the environment? Aluminum pots and pans? Smoking or not smoking? Drinking or not? Blows to the head? Lack of mental exercise? Depression? Too much education?

Rugg met Crome two years ago, while Rugg was working on

the Voynich, and proposed verifying Alzheimer's studies to suggest new approaches. Crome realized he was being handed a gift. In addition to becoming too specialized, science has become too dense. The National Institutes of Health's PubMed database alone lists more than 40,000 papers pertaining to Alzheimer's. Regardless of the scientific field—medicine, nuclear physics, mathematics, whatever—PhD students typically take a year just to read what others have done before them, and even then they are usually just scratching the surface. The papers pile up faster than you can turn pages. "I can't stay current in my own area—drugs—let alone all of Alzheimer's," Crome says.

So how does Rugg imagine he can tackle this vast field? He's not a doctor, and he freely admits that the sum of his Alzheimer's knowledge has come from the popular press. "The idea is, sit down with these experts, have a cup of coffee, and ask, 'Now, off the record, what are the main theories in your field?'" he says. "Then look at only those key papers in those fields. We would have consultants in each discipline and get them to help us crawl over those papers."

Rugg doesn't expect to find evidence of pattern-matching in a printed document, but there are always clues. "You look at the language," he explains. "Sometimes you'll see areas where the authors write, 'This result strongly resembles something else.' That's an opportunity to investigate more closely." Such a phrase is usually built on mountains of assumptions, which an author presumes (sometimes erroneously) that all potential readers understand.

As a first step, Rugg is preparing to apply for a grant of about $350,000—enough to pay for a postdoc and some expenses for three years. For the moment, he's resisted the temptation to look up a single Alzheimer's paper. "I want to approach this material fresh," Rugg says.

He's eager to start on the project because medical problems are so urgent. "Alzheimer's is a terrible way to die," he says. "If you could do something to help, you could feel really, really good."

Already there are signs that the Voynich breakthrough has

vanquished the nagging Catch-22. Following Rugg's success with the manuscript, potential collaborators seem to surface daily. Astrophysicists at Keele have agreed to work with him on sorting out seemingly incompatible theories in modern physics. A biochemist at UC San Diego approached him about untangling questions in exobiology, the study of the chemical origins of life. A researcher at the UK's University of Warwick wants to apply the verifier method to attention deficit-hyperactivity disorder. And an emeritus art historian at Rensselaer Polytechnic Institute thinks Rugg can help her resolve mysteries about ancient Greco-Roman settlements.

All these collaborations are in the earliest stages. "It could be that I just got lucky once and never again," Rugg says, then pauses. "But suppose we find a breakthrough in Alzheimer's and physics?" At the very least, he'll need a bigger kitchen table.

Rugg's team has since applied the Verifier Method to the study of autism. His position on the Voynich manuscript remains controversial. Rugg is author of a book about expert errors, entitled Blind Spot *(HarperOne Books).*

LAST OF THE BIG BANG MAN

1999

It's a scientist's worst nightmare. You make a significant contribution to one of the world's most important scientific theories —and everyone forgets you.

No one ever recognizes him, although he is arguably one of the most important scientists of the century. He just seems to blend into matter and light. On campus he's the predictable physics prof, emerging from the science building at Union College with his hands deep in his pockets, a suspender peeking out from under his tan sweater. But you can blow his cover with a single question: Where did we come from? Ralph Alpher knows the answer.

Back in 1948, Alpher submitted a PhD dissertation that gave birth to the scientific theory known as the Big Bang. He suggested, mathematically at least, how the universe could have begun in a superhot explosion fourteen billion years ago. The Big Bang. A few months later, he showed how to prove it. But in 1948, good math or not, these were loony ideas, and radio astronomy was a very young science. No one seemed willing or able to point a radio telescope toward deep space to confirm them. The years rolled by and everyone forgot about Ralph Alpher.

Then one day in 1964, two radio astronomers from Bell Labs stumbled upon the evidence that Alpher was right. Except they had never heard of him either. They got the Nobel Prize; he got bupkis. But that's exactly the kind of injustice Ralph Alpher is used to.

Roll the tape back to 1937. The kindly old physics professor was once a husky sixteen-year-old prodigy with dark hair and glasses. He received a letter from the Massachusetts Institute of Technology. It invited him to attend the school for free, on a full scholarship.

But there was a catch. The scholarship was good only if Alpher attended full time and did not work.

This was the Great Depression. Alpher's immigrant father was a home builder in Washington, DC at a time when no one could afford to buy a house. Alpher didn't even have train fare to Boston. How could he go to school if he couldn't work part-time for books and dining hall meals?

The letter told him to meet with an alumnus in Washington. Alpher went and talked to the alum for hours, hoping to find a way to make this work. But the guy kept turning the conversation back to the same subject—religion—and asked Alpher about his religious beliefs. "I told him I was Jewish," Alpher says. Soon after, a second letter came. The scholarship was withdrawn, without explanation.

"My brother had told me not to get my hopes up," Alpher says, "and he was damn right. It was a searing experience. He said it was unrealistic to think that a Jew could go anywhere back then. I don't know if you know what it was like for Jews before World War II. It was terrible."

When Ralph Alpher says something is terrible, you believe it. The word turns to ash as it drops from his lips.

Of course, Alpher had earned that scholarship. Just as he had earned the right to get credit for the Big Bang theory. That's what really drives him crazy: credit is everything to a scientist. "The important thing is to get the credit in the literature," he says.

"There are two reasons you do science. One is an altruistic feeling that maybe you can contribute to mankind's store of knowledge about the world. The other and more personal thing is you want the approbation of your peers. Pure and simple."

Yes, this has happened before, scientists who made good and never collected. Perhaps the most famous example is Gregor Mendel, the Austrian botanist and monk who tinkered with sweet peas in his monastery courtyard. He summed up everything he had observed about the propagation of genetic traits in two papers that were published without much notice well before he died in 1884. Only after the chromosome theory of heredity was nailed down in the early 1900s was he hailed as the father of genetics.

Alpher may be the Mendel of the modern age. "I don't really know Mr. Alpher," says Hans Bethe, ninety-three, Nobel laureate and often lauded as the world's greatest living physicist. But the theorist who helped Oppenheimer build the first atomic bomb and later devoted his life to disarmament does know Alpher's story. "I think it is a fact that he has not been given proper credit and it is a fact that he *deserves* a lot of credit."

Growing up, Ralph Asher Alpher had a code. It's in the Boy Scout Handbook: A scout works to pay his way. A Scout is true. That's Alpher. He rises formally and gives you a two-handed shake. He doesn't laugh; he chortles politely. He has worked steadily since he was twelve years old, a stagehand earning 50 cents an hour. Some weeks it was more than his father made.

Alpha is A, the first letter of the Greek alphabet. *Aleph* in Hebrew, *alif* in Arabic. It's a name that signifies beginnings—genesis, how the world began. Alpher had already thought about this by the time he was eleven. In Hebrew school he read Genesis and began to argue with his rabbi. Alpher believed his private stash of books told a more rational tale for the origins of the world and the organisms on it. The books were written by science heavyweights: Sir Arthur Eddington. Paul deKruif. Sir James Jeans.

The rabbi's book was written by the finger of God.

"It got hot and heavy," Alpher recalls. "I'd bring in quotations, with references, and he didn't want to have any part of it. Finally he said, 'You gotta go through the bar mitzvah for your father's sake.' Honor your father. 'And after that,' he said, 'I don't care what the hell you do.'"

Alpher did his duty. The bar mitzvah went well. When his mother, Rose, died of stomach cancer in 1938, Alpher was only seventeen but he did his duty again. Twice a day for eleven months he and his brother Robert attended services, morning and night, at The Hebrew Home for the Aged, where they could be assured of a steady supply of ten men for the *minyan*. Each morning they nibbled small egg biscuits, *kuchen*, and downed shots of Four Roses whiskey. Alpher did not care for alcohol. He was still in high school and working as a secretary to make money to go to college. Each morning he went off to take dictation with cheap whiskey on his breath.

Many years later, his father, Samuel, remarried. The new wife trashed Alpher's Eagle Scout merit badges. His uniform. His guide books. All of it gone. This happened when he was a much older man, with a wife and children of his own. But he never forgave her.

He searches his memory but he cannot recall the sound of his mother's voice, nor a single significant remark she or his father made to him. "No words of wisdom from either of them," he says. "Terrible, isn't it?"

He worked by day, took classes at George Washington University by night, at first in chemistry. Friends warned Alpher there were no jobs for Jews in chemistry. On to physics, then. Einstein's field. At the Naval Ordnance Laboratory, he worked on technologies to protect ships from magnetic mines. At Johns Hopkins Applied Physics Laboratory, he worked on torpedo exploder devices and guided missiles.

His thesis advisor was a hulking Soviet defector named Gamow (pronounced GAM-off). A huge man whose idea of dinner was a few gin martinis. In his spare time, Gamow wrote the kinds of popular science books Alpher used to read as a kid. Gamow tossed Alpher a dissertation problem dear to his own heart: the origin of the elements. Gamow was no doubt influenced by the ideas of earlier scientists working in the 1920s, namely Alexander Friedmann, a Russian whom Gamow had studied under, and Georges Lemaitre, a Belgian priest, who, independent of Friedmann, proposed that the universe must have begun as a "cosmic egg," or primordial atom. Gamow was interested in the mechanics. If indeed the universe had a starting point, as the earlier scientists had suggested, how did it actually happen? Gamow hadn't hammered out any sort of theory. That became Alpher's task.

Alpher was excited. Despite the work of an American, Edwin Hubble, who showed that the universe was indeed expanding in a manner consistent with the ideas of Friedmann and Lemaitre, the origin of the universe was off the beaten path for physicists then, weird science. The closer you got to Time = 0, the more the mathematics seemed to self-destruct and the more impossible it became to determine or predict how the particles behaved. Gamow said, hey, forget about the exact beginning of time. Everyday physics had to kick in at some point. *Our* physics. Let's start there, Ralph.

Alpher's signature skill was apparent from the outset. Eamon Harper, a GWU physicist and science historian, has spent three years researching Gamow's biography. Sifting through papers at the Library of Congress, Harper noticed that a picture of Alpher as detail man emerged. "Alpher was very meticulous. Even in their letters, he would write detailed calculations. Gamow would make calculations too, but he was always thinking of the quickest way to get to the end, to get to the essence of the thing."

Alpher and Gamow focused on the point when the universe had cooled to a state consisting of radiation and matter, *ylem*, a Greek word for the primordial stuff of life. Alpher's final draft gave the mechanics for what happened next: Ylem began as a cloud of neutrons, neutral particles. Some of these decayed radioactively, forming protons, electrons, neutrinos, the building blocks of matter. As the universe cooled and expanded, Alpher wrote, the remaining neutrons, plus protons and electrons, combined to form all the elements.

It later turned out that this picture was not entirely accurate.

For one thing, the process apparently stopped with elements that have an atomic number of 5 or higher. But that's not the point, Harper says. For the first time, a workable mathematical formula had spelled out how the universe had been born. "It's hard for laymen, even scientists, today to realize how visionary their work was," he says. "The whole idea of suggesting that you can, on the basis of science, explain how the distribution of elements—the material we're all made of, how we came to be—that was not a suitable question for scientists. It was mystical, theological."

But this time, God wasn't in the details, Alpher was.

As he was finishing up the paper, Alpher came down with the mumps. This was how the last draft of his Big Bang paper was born: a swollen-jawed doctoral student sitting up in bed, writing by hand, passing papers to his young wife Louise, who sat nearby, muttering and typing. Her husband had spent six years in night school getting his bachelor's degree. Add two years for the master's, three for the doctorate. Look at him, she thought. He should be resting. The couple had met over a bridge game at GWU in 1940. At the time, Louise was a psychology major, attending night school too, working days as a secretary in the State Department. They were married two years later, a month after Pearl Harbor. The rabbi was Reform; his poodle came to the ceremony. Alpher's Orthodox family did not.

Louise typed the paper and Alpher presented it to George Gamow, a guy who had once unsuccessfully tried to defect from the Soviet Union by paddling across the Black Sea to Turkey. Gamow's approach to physics was no different: look for shortcuts. Think big. Have fun. Today they call Gamow brilliant; back then he was eccentric, as Alpher was about to find out.

Gamow excitedly flipped through Alpher's paper and grinned.

"Now I want to do something I've always wanted to do."

"What's that?"

"I want to put Hans Bethe's name on it." Gamow was secretly in love with the idea of playing off the authors' names—Alpha, Beta, Gamma.

"The hell you are!" said Alpher.

This is my dissertation, he said. How can Gamow kid around like this? As it was, Alpher was worried that the paper's topic was so speculative that journal editors would be inclined to reject it anyway. Why kid around? Bethe, known for his analysis of energy production in the sun, had worked on the Manhattan Project at Los Alamos. He hadn't done of a lick of work on this paper. But Gamow persisted; Alpher inexplicably relented. Off went the paper to a physics journal.

The day they dropped it in the mail, Gamow and another scientist—Robert Herman, whom Alpher had met at Johns Hopkins—appeared in Alpher's office with a bottle of Cointreau whose label had been doctored to read YLEM. Alpher hated the orange liqueur but downed a ceremonial shot just the same. (The bottle is now on display at the Smithsonian's National Air & Space Museum.) Some time later, Alpher got a card from the prestigious physics journal *Physical Review*, and shuddered. Publication was set for April 1, 1948.

Publication made the "Alpha Beta Gamma" paper hot. But Alpher still had to defend it before the faculty to get his PhD. When the day arrived, he nervously pulled on his academic robes and entered the auditorium. He froze. The place was packed with three hundred people. Newspaper reporters. Science writers. Bethe himself was there, as were other top scientists such as Ugo Fano and Charles Critchfield. Typically, there were rarely more than a dozen people attending oral exams. But hundreds of people had come to see the kid who thought he knew how the universe was born. Alpher marched behind Gamow and into the ring to take the committee's questions. Asked how long the entire process of primordial nucleosynthesis had taken, Alpher answered 300 seconds. The press went crazy. "Scientist says world was created in five minutes," read a skeptical Herblock editorial cartoon in the next edition of the *Washington Post.*

"I began to get letters from people saying novenas to save my soul because I had dared to trample on their concept of Genesis," Alpher says.

A few weeks later, at his home on Farragut Street, home builder Samuel Alpher settled in with his favorite newspaper,

Der Tog. And gasped. Staring back at him above the fold of the Yiddish weekly was the smiling face of his son. He read the story a few times but didn't really understand it. But it didn't matter. His boy Ralph was on the front page of *Der Tog*.

Alpher followed up his original paper by publishing eighteen research papers with his new friend and colleague, Robert Herman, a wiry New Yorker with an elegant mustache. Alpher and Herman's first paper predicted what astronomers could find in space to prove the Big Bang actually happened. The paper said, Look, the radiation from the primeval explosion still exists. Some of the heat is still out there. It's with us now. It's been bouncing around in space for fourteen billion years and has cooled to a temperature of 5 degrees Kelvin. Which is still plenty cold, about 450 degrees below zero Fahrenheit.

Today that calculation seems very simple. Just a proportion: Radiation seconds after the Big Bang divided by the amount of Matter then, equals Radiation now divided by Matter now. The formula was small enough to write on the back of an envelope. But try coming up with it in 1948. No 800-megahertz laptops. No e-mails. Not even a graphing calculator. Alpher and Herman had access to one calculating machine. Once they started it up, it clacked so badly that they had to leave the room. It took the two men all summer to confirm their "simple" formula.

It came down to this: if you found the radiation, you proved the Big Bang theory. But in 1948 astrophysicists told Alpher and Herman that there was no way to measure this so-called background radiation in space. You couldn't just go out and wave around an antenna. You needed to train a specific receiver on a specific band of the electromagnetic spectrum. Alpher and Herman couldn't do it themselves; they weren't radio astronomers.

They didn't give up. They gave talks. Alpher published more papers with Herman and Gamow. They buttonholed radio astronomers wherever they went. At one point, they even held a press conference, trying to drum up interest in their theory. Nonetheless, all their effort resulted in a cosmic goose egg. So they let it drop. Years later, when younger scientists would say that their work had been forgotten, Alpher would think: How could that be, after all this? The young guns would say Alpher and Herman hadn't tried hard enough, and Alpher would say, "We were there, they weren't." Some scientists would even suggest that perhaps Alpher and Herman didn't know that the 5 degrees K was *microwave* radiation, an insult that would really grate on Alpher: "A sophomore physics student would know that."

The two men had growing families to support and were unhappy at John Hopkins. In 1955, Alpher headed for General Electric in Schenectady with his wife, son, and daughter. He worked on color television, energy conversion, liquid metals, gas dynamics, and wrote about a hundred scientific papers for GE alone. Herman went to General Motors. They stayed in touch, but wrote only four more papers together. "We slacked off, rightly or wrong, after 1955," Alpher says.

The world turned around. In 1965, the *Astrophysical Journal* hit Alpher's desk, featuring two articles, a scientific double whammy in plain paper wrappers:

Item 1: Arno Penzias and Robert Wilson, two radio astronomers using an ultrasensitive radio telescope at Bell Labs in Holmdel, New Jersey, unexpectedly detected unwavering radiation of 3.5 degrees K bathing the universe.

Item 2: Working independently, a four-man team, led by physicist Robert Dicke at Princeton University, pegged the finding as radiation left over from a primordial fireball. The team had predicted heat of 10 degrees K and were building a telescope in order to measure it when Penzias and Wilson scooped them.

Alpher felt ecstatic for about a minute. Then his heart was in his throat. He paged through the two articles looking for his name. He found a single line indicating that in the 1940s, he, Herman, and Gamow had envisioned a nucleosynthesis process like the one mention in the report. But there was not a single mention of Alpher and Herman's 1948 5 degrees K prediction. This struck Alpher as weird, because several months before, the editor of the *Physical Review* had actually sent a paper from the Princeton team to Alpher and Herman, asking that they review it, a common practice at scientific journals. The two men had told the editor that the Princeton team had duplicated their work. They suggested rejecting the paper on these grounds. The editor had later sent a revised version of the paper to Alpher and Herman. It still didn't credit them. Alpher and Herman had sent the revision back, this time citing exact references. Nothing happened. Months later, the Princeton paper and the Bell Labs paper had now appeared together in a *different* journal.

Alpher was appalled. Why hadn't the scientists credited him and Herman for the original prediction? That was the way the game was played, the way the men of science did things. Where was *their* code?

Alpher pauses now in his tale. It's midday now, and he's been talking since morning. Sitting in the cafeteria at Union College, his academic home for eleven years, Alpher removes his glasses and rubs his eyes. Around him, midterm-crazed undergrads drink coffee. Alpher's bassoon voice drops. "People were wondering why we were upset, but they never sat in our shoes," he says. "Was I hurt? Yes! How the hell did they think I'd feel?"

Left out of the glory, Alpher and Herman and Gamow hit their typewriters in 1965 and never stopped. The stream of print was punctuated by several events. Gamow's death in 1968, for one. In 1971, James Peebles, the key author of the controversial Princeton paper and today Albert Einstein Professor of Science at the university, set the record straight in his book *Physical Cosmology*. But Alpher and Herman kept writing letters. Stephen Hawking got one in the late 1980s, after he credited the Princeton team with the critical prediction in his bestseller, *A Brief History of Time*. (The passage remains unchanged to this day.) Another letter to Penzias addressed Alpher and Herman's "high frustration level," claiming falsehoods had become "widely entrenched" in the literature. They said they had dealt with the matter in a "gentlemanly way." And they enlisted Penzias's help to set the record straight "in the best traditions of scientific integrity without embarrassment to anyone." The words were pure Alpher: proper, but plenty angry.

The next insult: In 1978, Penzias and Wilson won the Nobel

Prize for physics. While Penzias was working on his acceptance speech, he invited Alpher, who was giving a lecture at nearby Rutgers, to come by for a visit. Penzias had, perhaps wisely, stayed out of the Princeton controversy. Now he generously planned to delineate the history of the origin of the elements, citing the work of Gamow, Alpher, and Herman in his acceptance speech.

The meeting was difficult for Alpher. "I spent a day and a half with him in which I gave him a crash course on cosmology, and he didn't know a damned thing." Later he says, "I was miffed at the time that they'd never even invited us down to see the damned radio telescope. It was silly to be annoyed, but I was."

A month later, Alpher suffered a heart attack. He blamed it partly on that visit. Penzias, now a consultant in San Francisco, says the news "really made me feel bad. [Ralph] has an enormous personal investment in this."

Until the heart attack, Alpher's children did not realize how much stress their father had endured in his crusade for recognition. Or even how important it was to him. Alpher had shared his concerns daily with Louise, who many times urged him to drop it, but he had deliberately shielded his children. His son was in his twenties before he ever learned of the magnitude of his father's accomplishments.

"He doesn't let you know when something is bothering him," says his daughter, Harriet Lebetkin, a piano teacher in Connecticut. "I was pregnant with my first child [when Alpher suffered the heart attack]. Here he was, getting ready to be a grandfather, and I almost lost him."

Eventually, the truth did sort itself out. Scientists today accept that the work of the trio contributed greatly to the overall development of the Big Bang model. "The work of Gamow, Alpher, and Herman placed Friedmann's and Lemaitre's conception of the universe on a material (and thus physically observable) foundation," Harper writes.

As a younger scientist working under the more seasoned, charismatic and better-known Gamow, Alpher's contributions seemed almost destined to be overlooked. (The comical byline on his first paper probably didn't help matters.) Nevertheless, in the end, Alpher and Herman received six major scientific awards for their achievements, among them a medal from the American Philosophical Society, the Magellanic Premium, and one presented to them by the King of Belgium. But the big one, the Nobel, was lost to them forever.

Alpher and Herman's story raises interesting issues about the personal side of science. Yes, all human beings have feelings. Yes, every person is allowed to reach their boiling point. Scientists just happen to belong to a profession where you are not allowed to show it.

"It was a horrible injustice but I don't know what you do in such a circumstance," says Vera Rubin, a friend to both the Alpher and Herman families, and an astronomer who received the National Medal of Science in 1993. "It would have been nice if they had had happier lives. They could have known that they did something very valuable, and they could have been happy with this. I think perhaps injustices are in the eye of the beholder, unfortunately." And then she says, "There's no doubt that they could have been and should have been treated nicer by the [Nobel] committee. They really do have a legitimate complaint, but they could have responded a little differently... They had not gotten the recognition they deserved, but if their personalities had been different, they could have been happy with the knowledge of this great thing they had figured out. And they perhaps could have even been treated better by the committee if they had not just been so obviously angry."

In November 1989, when NASA launched the Cosmic Background Explorer, or COBE (pronounced KO-bee), a $150-million, two-and-a-half ton satellite designed to investigate the questions raised in Alpher and Herman's papers, the two scientists got to see it go up as guests of NASA. COBE subsequently detected cosmic background radiation at 2.7 degrees K. "I think maybe a couple of thousand papers have been written since COBE measured it," says John Mather, then COBE project scientist. "It's really incredible. There are more papers than there are pieces of information."

Herman died in 1997. That left Alpher as the last of the Big Bang theorists. Today, in his daily life, Alpher is trying to leave the past behind: "These things are of no consequence anymore," he says. He has thrown himself into a stream of local activities: tutoring kids, mentoring scouts, working for the local public TV station. He's busy trying to finish a book about cosmology that he began seven years ago.

Not long ago Alpher got an invitation to speak in Odessa, Ukraine—the hometown of both his father and his mentor, Gamow. He declined. Louise has been ill. "It's been a rocky road," he says of their fifty-seven years together as husband and wife. "We had our ups and downs. It hasn't been all smooth. Physicists as a class are a peculiar bunch. Our primary interest is in science and in the work, and sometimes affairs at home take second place. And I'm sure that hurt Louise along the way very much. But we survived."

As a boy Alpher often sneaked off to peek through a telescope at the US Naval Observatory on public viewing nights. Today, in his spare time, he is the administrator of the Dudley Observatory and its library, housed in a local senior center. Twice a week he heads over, fishes for the keys, unlocks the door, hits the light. Inside are marble busts, rare books, and an antique telescope. "Here's what I do," he says. He hunches over a small Radio Shack answering machine. "I hit the button and I follow the directions," he jokes. There is a beep, and he begins to read from a freshly printed script: "If we're lucky enough to have clear skies this weekend, you won't even need a telescope to enjoy the best show around. On Friday night the crescent moon, Saturn, and Venus will be cheek by jowl in the western sky at dusk..."

On he goes for several minutes, reading all the information amateur stargazers will need to have a wonderful weekend. Then he hits the button. "Now I check it," he says, and lifts the receiver on a second phone line. He dials the number of the observatory's Skywatch phone line and waits. The first phone rings. The machine clicks. A booming voice flows from the receiver in his hand. Alpher's voice. The last of the Big Bang men nods: Not bad, huh?

. . .

Ralph Alpher died in August 2007. Shortly before his death, he enjoyed greater recognition for his contributions. Reporters and TV news crews from all over the world flocked to his home to interview him for news segments and short documentaries. His life was adapted into a play. Oxford University Press published his book, Genesis of the Big Bang. *Science writer Simon Singh unhesitatingly hailed him "the forgotten father of the Big Bang." And perhaps most meaningfully, Alpher was awarded the National Medal of Science by President George W. Bush.*

WHERE CHIMPS GO TO DIE

2002

What happens to lab chimps when scientists don't need them anymore? They don't just go away, and you can't release them in the wild. If they're lucky, they find a home among humans who care.

I was interviewing a scientist one day when he began griping about how difficult it is to get a chimpanzee for medical research. "They're expensive," he said, "and you've got to pay all this money into a social security plan to take care of them when they retire."

Retired chimps? I thought. Just where do they go to retire—and what do they do when they get there? Eat bananas? Play shuffleboard? "I heard they put them on an island sanctuary in Liberia," the researcher said. He didn't know much more than that, and of course it wasn't his job to know. He is the scientist. He uses chimps to answer scientific questions. Someone else deals with what comes afterward.

For a while I thought about going to Africa to look for that chimp island. In my mind's eye I envisioned a paradise where repatriated American chimps lived free of humans, free of the cages that once confined them. But that turned out to be a pleasant fiction, a tale told to lab workers foolish enough to ask. By then I had observed chimps in zoos, read about them in the scientific literature, and immersed myself in a world of animal sanctuaries that is stranger, more interesting, and more disturbing than I could have imagined.

Eighteen chimps do live on a pair of islands in Liberia, most of which were bred on-site by an American hepatitis research laboratory. And sanctuaries throughout Africa shield wild chimps from poachers. None of these places will accept US-bred lab chimps after we're done with them. That's why sanctuaries are needed in the US.

At this moment, the United States is up to its ears in chimps. During the 1980s, laboratory supply companies bred chimps like crazy to meet the demands of AIDS and hepatitis researchers. That didn't work out too well. By the late 1990s, researchers conceded that while some chimps become HIV-positive, almost none develop full-blown AIDS. At least two hundred chimps have been exposed to HIV, yet only two may have died of AIDS. The researchers switched to macaque monkeys. For a short time the National Institutes of Health, which funds much of the biomedical research in this country, considered euthanizing HIV-exposed chimps when they were no longer useful. The NIH later decided not to, in part because the animals are listed as an endangered species. But the surplus has mounted—today more than 1,600 live in various primate facilities in the United States—and humans have begun to ask themselves a serious question: What are we going to do with these animals?

During his last weeks in office, President Clinton signed the Chimpanzee Health Improvement, Maintenance, and Protection (CHIMP) Act, which mandates a national system of sanctuaries for chimps who qualify, but it is likely to be two years before any new refuges are ready. In the meantime many animals will have to remain in labs. So far, about two hundred chimps have been earmarked for retirement. When chimps do enter sanctuaries, there will be a string attached: If a sanctuary owner takes government money, he or she must be prepared to send the chimps back to the lab for further research if asked to do so—a stipulation that infuriates those who believe that plucking a chimp out of retirement negates the concept of sanctuary.

IN 1997, NEW YORK UNIVERSITY DECIDED TO GET OUT of the chimp business and shut down its Laboratory for Experimental Medicine and Surgery in Primates. It's a common story. Around the world, chimp labs are dwindling: The last remaining facility in Europe—in the Netherlands, with 105 chimps—is closing. New Zealand has banned chimp research, and in the United Kingdom no new licenses are being granted for this work. The only other nations that still use chimps for medical research are Japan (370 chimps), Liberia (18 chimps), and Gabon (72 chimps).

When labs close, chimps are up for grabs. Of the 250 chimps who "contributed," as they say in lab parlance, to experiments at the New York University lab, ninety were placed in sanctuaries; the rest were transferred to other labs. The most difficult to place were those that had been exposed to HIV or hepatitis, both transmissible to humans.

Later in 1997, the Fauna Foundation outside Montreal became the first sanctuary in North America to give retired, HIV-exposed chimps a home. Fauna had been an animal refuge for close to ten years, and the chimps joined a menagerie that included goats, pigs, chickens, rabbits, horses, turkeys, geese and ducks, cats and dogs, Scottish Highland steers, cows, llamas, emus, rheas, capuchin monkeys, one guanaco, one Jacob's sheep, one ostrich, and one donkey.

Until Gloria Grow, forty-six, the owner of the sanctuary, and her husband, Richard Allan, forty-nine, announced their intention to shelter fifteen of the chimps from the New York University lab, eight of them HIV-infected, the refuge had been regarded by residents of Chambly as a harmless oddity. Suddenly, the local planning board challenged every variance Fauna requested. When the board saw plans for an elaborate, secure chimp house complete with cages, they acquiesced. Nonetheless, teachers conducted chimp drills at the elementary school, instructing kids to hide in the classroom closet if a chimp appeared in their midst; and police laid in a store of Tyvek suits and tranquilizer guns.

Fauna's chimps live in a nine-thousand-square-foot building that looks a bit like a day care center. Despite the cages, living conditions here beat lab life knuckles down. The outdoor play area contains picnic tables, chairs, and swings; indoors, two-story playrooms are packed with toys, blankets, and more swings. Chimps can also rest in private cages that give them access to the indoor play space but keep them separated from humans. They can snack on fresh fruit and vegetables, or page languidly through Victoria's Secret catalogs. The human form enthralls great apes.

Each day Grow and her staff of three whip up three savory meals. The menu includes fruit, oatmeal, spaghetti, potatoes, soups, stews, steamed vegetables, and rice, as well as the occasional vegetarian pizza and birthday cake. Staffers mix gallons of concentrated orange juice every day and laboriously pour it into empty water bottles with plastic lids. "Most of the chimps know how to unscrew them," an employee explains, "but sometimes they like to stick a canine in the cap to bite it off."

One day in October 2001, the chimps are lounging around after lunch, picking at their food, grooming themselves, and playing with toys. Some snatch plastic cups of hot Tetley tea off the trolleys parked in front of their enclosure, sip carefully, and return the cups through the bars without spilling a drop. Grow brushes Tom's back with a small brush. "Let me see your fingernails," she says. Tom holds them up for inspection. Another chimp, forty-two-year-old Annie, the oldest and a surrogate mother to the others, spots the brush and gestures through the cage for it. "You want the brush?" Grow asks as she slips it to her. Annie spends a couple of blissful minutes stroking her coat.

A few minutes spent watching chimps manipulate objects like cups, bottles, and brushes quickly demonstrates why biologists regard them as the top tool users in the animal kingdom, after us. Besides being dexterous, they are intelligent, strong, and often aggressive, especially as they grow older. Chimps have a sense of humor, which any visitor to Fauna notices immediately: They delight in teasing humans as well as each other. They routinely spit water at their caretakers, cleverly varying the pattern to confuse them. They also understand and respect social hierarchies: A beta male accepts his lot when an alpha male swipes his orange but goes ballistic when a lower-ranking female does the same.

The chimpanzee's ability to learn can be humbling. In 1967, psychologist Roger Fouts taught chimps to use American Sign Language, which they mastered and taught to other chimps. Since 1983, psychologist Sarah Boysen has been teaching chimps at Ohio State University to do simple arithmetic; in 1991 she figured out how to teach them fractions. In 1999, a landmark paper written by Jane Goodall and eight other prominent primatologists asserted that chimps use their smarts to master their environment. Specifically, the primatologists demonstrated that chimps can codify cultural behavior—how to hunt, how to eat ants, how to groom oneself and others—and pass that knowledge along to their young ones. Chimps that live in the Gombe forests of Tanzania have been observed dancing, apparently to make the rain stop.

In the wild, bands of chimps will rove the jungle for six or seven miles a day, joining together to hunt monkeys, which they delightedly eat, usually after dashing out the smaller creatures' brains. That's the side of chimps humans rarely see or choose not to see. However, chimps can also be kind. A chimp was once observed trying to help a wounded bird to fly at a zoo in England.

At Fauna, there is constant physical contact between humans and chimps. The tiniest scrape or scab on Grow's hand will elicit concern from a chimp, usually in the form of a kiss. According to the standards of the Centers for Disease Control, Fauna's chimp house is a biohazard facility. If this were a US lab, workers would be required to wear Tyvek suits, goggles, masks, or hairnets, mandatory garb worn by researchers studying HIV or hepatitis. But Grow and her workers wear street clothes, unless a chimp has an open wound or needs surgery, in which case they follow aseptic procedures, donning gowns and gloves just as one would with a human patient. They believe that if the animals are treated well, they will not harm their caretakers. The theory has proved true so far, but the chimps do get into fights with each other that require bandages or surgery.

Now Grow calls to another animal: "Billy Jo, is your show on?" She peeks at the TV. "Oh, it's Rosie. Don't worry, Oprah will be on soon." Grow agonizes about keeping the chimps stimulated. Because they remain caged, she wants to help entertain their restless spirits. Hence the painstaking preparation of meals, classical music piped over the stereo, hanging spider plants, brushes and paints, Halloween decorations, Christmas lights, birthday parties, crackling fires in the wood-burning stove, strands of red licorice, and aromatherapy candles.

In the United States, lab animals fall under the jurisdiction of the Department of Agriculture, which inspects labs and enforces the Animal Welfare Act of 1985. By these standards, the Fauna chimps were treated well in their former lives: They got adequate food and shelter, their cages were clean, and they received the occasional toy or orange. But Grow and others like her consider those standards weak and seek to do better. "I want them to be happy," Grow says. "To treat chimps well, you should treat them as you would victimized people. Because they have been victimized. Terribly. Oh, it's so awful what they've been through."

Annie, for example, was born in Africa, probably in 1959, then captured and brought to the United States. She gave her life to humans for more than thirty-five years—at least fifteen in the circus, followed by twenty-one in the lab as a breeder. When she refused to mate, she was artificially inseminated. Her child was transferred to another facility at age three. Another Fauna chimp, Rachel, was born at the Institute for Primate Studies in Norman, Oklahoma, in 1982. Rachel was sold for $10,000 as a pet but ended up at the New York University lab when her owners divorced. Rachel, who had grown up taking bubble baths and prancing around in dresses, spent the next eleven years isolated in a cage. Today she occasionally bursts into screaming and scratching fits, lashing out at her own hand, apparently because she thinks it is attacking her. Her outbursts have diminished somewhat since she arrived at Fauna in 1997, but her body is still covered with self-inflicted sores.

"Jeannie was going to be euthanized—did you meet her?" Grow asks. Lab workers had to medicate Jeannie, an HIV-exposed chimp, to stop her seizures, during which she ripped out her fingernails and thrashed any human or chimp nearby. "She had a nervous breakdown before she came here, but she has made lots of strides in her development. They all have. They put on weight, grow more hair, their coats are shinier. They sleep better at night. We don't have nearly as many fights as we used to when they first came, and they have learned to vocalize more like real chimps."

Fauna's annual budget for the entire farm is $60,000. The food bill is $40,000; the rest covers medicine and necessities such as bedding straw, tools, and equipment repairs. In a good year, $15,000 of that comes from private donations. Fauna is not eligible for funding under the CHIMP Act because the sanctuary is in Canada. Grow says she would not apply for funds even if Fauna were eligible because of the requirement that sanctuary owners return chimps to labs on demand: "I would never send them back. Who would?"

The bulk of Fauna's operating revenue comes from a dog grooming business and from Allan's veterinary clinic. On the first night of my visit, Allan, a French-Canadian who has been doctoring animals on the outskirts of Montreal for twenty-seven years, arrived at dinner in scrubs, looking exhausted. Delighted that Fauna would be the subject of this article, he quipped, "Tell them we need money."

One recent cold and wet morning, Grow is chopping fresh vegetables for the rabbits and pigs when her sister Dawna Smith, who works in the chimp house, chugs up to the barn in a Volkswagen. "Get in," she calls to Grow. "I need you to come look at Pablo."

"What's wrong with him? He was fine last night."

"In."

Up at the chimp house, the thirty-year-old, almost two-hundred-pound chimp struggles to make himself comfortable in his nest—a pile of blankets on a twelve-foot-high platform inside the chimp house. He can find no peace. First he sits, then he stands, repeating the process over and over: sitting, standing, sitting, standing. He wheezes constantly.

Since the day of his arrival five years ago, Pablo has been ill. One winter he developed a cough that X-rays showed to be bronchitis. Medication helped, but each autumn Grow worries that Pablo's cough will return. Still, she has never seen him behave like this.

She dashes up a spiral staircase to offer him more blankets, an antibiotic, and a Tylenol. The big-lipped chimp graciously accepts the blankets but spits out the pills. Grow runs out to find her husband, who is busy spreading a load of red cedar mulch someone has just donated to the farm. Allan's practice, largely cats and dogs, did not prepare him for the variety of animals with which he now shares his land and board. To prepare for the chimps' arrival, he spent a few days training at the New York University lab with veterinarian James Mahoney.

"What do you think is wrong with him?" Grow asks.

"He's dying," Allan says, staring into the cage.

Grow doesn't want to hear that. Her husband often acts on the premise that there is only so much one can do for an animal, especially a wild animal who will not permit a detailed physical exam. But Grow was raised to believe that she should go to extremes to help sick animals. Her father, an electrician, thought nothing of stopping his truck in heavy traffic to rescue a wounded seagull. Now, as Grow looks on, Allan phones Mahoney and leaves a message on his voice mail, then heads off to resume mulching.

Hours tick by. At lunch, Grow and her sister, employees, and volunteers sip soup and munch eggplant casserole in silence. When Allan comes in to wash up, Grow asks, "What do you think we should do?"

"What should we do?" Allan repeats. "We should wait and see how he feels tomorrow."

"Wait and see? If I were one of your patients, you think I'd want to hear that?"

"What do you want to do?" says Allan. "Tranquilize him?"

When Allan heads back outside after lunch, Grow asks him again what he thinks is wrong. He repeats the two words he uttered that morning. The words stab the air, and then he is gone. Grow is left to ponder their meaning in a congress of women. Pablo can't be deathly ill, she and her staffers decide. He is only thirty; captive chimps can live to be sixty. Her sister dissents. "The thing is," Smith says in a measured tone, "Richard's always right."

Mahoney phones at 2:00 p.m. The big chimp is down, still breathing hard. Earlier, he drank some juice with antibiotics but vomited it up. Mahoney offers possible diagnoses: pneumonia, a cardiac problem, a twisted intestine. Given Pablo's past, pneumonia seems likely. Allan is instructed to administer three injections, one after the other: an antibiotic, a diuretic, and cortisone for shock. If Pablo has pneumonia, he should feel better after the first shot. Allan drops the phone and dashes to get his bag.

In labs, monkeys and chimps are trained to present their arms for blood draws. Pablo had always resisted, so he was usually tranquilized—"knocked down," as the lab techs say, with a dart fired from an air pistol. Allan fears he'll have to break out the darts for the first time ever at Fauna. When Pablo sees the needle, he thrusts his arm out. Allan is stunned. "This guy never liked needles, but he gave me his arm. Didn't put up a fight."

Minutes after the injections, Pablo lies back and closes his eyes. His face is immobile; a black arm hangs limply off the side of his nest. Allan carefully unlocks the gate to the chimp enclosure, and Grow rushes up a ladder. She grabs Pablo's hand and feels a twitch. Life shudders out of the great ape's body. Grow begins to cry but manages to help carry the body to the floor. Allan confirms the chimp is dead, and Grow insists that the humans leave the enclosure to allow the other chimps a chance to see Pablo.

Normally, when a lab chimp dies, he dies alone in a cage and is whisked away. Grow believes chimps should be allowed to witness everything. A couple of times Allan has performed surgery in the kitchen in front of the cages, where all the chimps could see him. "Someday, when I die," Grow says, "I want to be placed right here where they can all see me and know that I am gone."

So, as Grow and her staff sit weeping outside the enclosure, the chimps approach Pablo. Alone or in pairs, they tug at his arms, open his eyes, groom him, rub his swollen belly. Annie pours a cup of juice in his ear. Grow says it might be an attempt to annoy Pablo and wake him up. Before long, the chimps wander off, hooting. The hoots blossom into screams, and soon the walls of the chimp house echo with the sound of knuckles pounding steel.

The night of Pablo's death, Allan conducts a hasty necropsy, but neither he nor his colleagues have had much experience handling a large and potentially infectious animal. His veterinary clinic is well-supplied with Tyvek suits and latex gloves but short on masks and goggles. Everything seems too small for Pablo's frame: the clinic's back door, the operating table, especially the freezer into which Allan and a sobbing Grow stuff his body when the procedure concludes.

Eighteen days later, after Grow has pleaded unsuccessfully with different agencies to perform an official necropsy, the Montreal health department presses a pathologist into service at the vet school in Saint-Hyacinthe. The immediate cause of death is listed as an acute lung infection, but the physician who examined the body also found an abdominal infection and mild hepatitis. Internally, the animal's organs were crisscrossed with thick, fibrous scars, most likely the remainders of various procedures. To do an animal biopsy, a technician uses a punch tool to clip out a chunk of tissue. The procedure leaves a large hole that, if infected, can take years to heal. Pablo was also vulnerable to infection on another front. Darts fired from an air pistol are, by definition, non-sterile; each penetration carries germs from the surface of an animal's skin into its body.

According to his research dossier, Pablo, known as Ch-377 at the New York University lab, had been darted 220 times, once accidentally in the lip. He had been subjected to twenty-eight liver, two bone-marrow, and two lymph-node biopsies. His body was injected four times with test vaccines, one of them known to be a hepatitis vaccine. In 1993, he was injected with 10,000 times the lethal dose of HIV. The barrel-chested chimp had shrugged off AIDS and kept hepatitis at bay only to die of an infection aggravated by years of darts, needles, and biopsies.

"We always knew the chimps had a lot of problems," Grow said two months after Pablo's death. "But we always thought they were problems we could take care of—because they were on the outside. Now we are learning that there are a lot of things going on inside them that we may never know about. Annie's sick now. Jeannie's sick now. What happened to Pablo wasn't unusual; it was average."

Activists insist that animal-free science is already here—in the form of in vitro research, data gleaned from autopsies, clinical observation, and epidemiology. But the scientists who work with chimps say that the inoculations, biopsies, and knock-downs, though regrettable, are necessary.

"I think the idea of moving to humans is nonsense," says Alfred Prince, the hematologist who heads chimpanzee research in Liberia. "Ethics committees in hospitals are getting tougher and tougher, and the work you can do in people is less and less. We will probably always need animal models...I think the answer is, if you're going to do this work on chimps, you better take really good care of them."

Other researchers, including primatologist Roger Fouts, believe that the days when we are willing to imperil an endangered species for our own sake may be numbered. Until then, research will proceed, and people like Gloria Grow will be left to deal with the results, as she did in January 2002, when Annie, the grande dame of the chimp house, died. Her body is awaiting a necropsy. Then the body will be sent, like Pablo's, to a local crematorium that donates its services to Fauna.

Grow plans to bury some of the ashes of both animals at the sanctuary. This spring Jane Goodall will take the rest of the ashes with her to Tanzania to sprinkle in the forests of Gombe, where chimps dance to stop the rain.

At this writing, six of the sanctuary's original residents have died, and those vacancies have been offered to new residents. Today, Fauna Foundation is home to thirteen former lab chimps.

THE MEDICAL DETECTIVE

2007

He answers his own phone, makes house calls, and runs his office without the help of nurses and assistants. Meet an unusual doctor who just happens to relish unusual medical mysteries.

A woman named Ethel Moore* arose one morning in the summer of 1999 and traveled seventy-six miles from her home in upstate New York to see a doctor in Manhattan. Moore was seventy-four years old, a stylish, humble woman who looked like someone's mom. Once in the doctor's office, she disrobed to reveal a secret that had both plagued and embarrassed her for two years and which she had kept hidden from friends and neighbors. A hairy tan-and-brown rash covered her body from the neck down. She had visited numerous doctors to rid herself of the affliction, but to no avail. The pathology lab of a prestigious Manhattan institution had examined biopsies of her lesions and pronounced them cancerous; an oncologist had dutifully prescribed chemotherapy. To Moore, that felt like a death sentence. She wanted a second opinion.

Her new doctor, Thomas Bolte, thought that she was more than entitled to one. As he flipped through Moore's voluminous medical files, he could see that something was destroying her immune system—but he did not know if it was cancer. As a young physician he had worked at a pathology lab, and he knew from experience that labs occasionally made regrettable errors. Under the microscope, a skin cell might look cancerous when it really wasn't. Histopathology, the microscopic analysis of diseased tissue, is more art than science. If the lab had gotten the diagnosis wrong, chemo might kill this patient. Bolte had never seen anything like this rash in his life, but he was determined to save Moore from it.

Bolte, who today is still boyish at forty-five and still working out of the small one-room office where he first examined Moore, is a doctor obsessed with mysteries. He calls himself an unusual symptoms investigator, his term for a doctor who picks up where other doctors fail. Patients find him on the Internet, and they appear in his consulting room like characters out of a foggy Sherlock Holmes story: The Case of the Migrainous Art Dealer. The Case of the Irritable College Grad. The Adventure of the Chemically Sensitive Sleeper. For fees that are astonishingly reasonable by New York standards, Bolte takes these patients under his wing and tries to ferret out the reasons for their illness. He usually succeeds.

"If a person calls our radio show and says they have a problem that no other doctor has been able to help them with, that they can't crack the case, I refer them to Bolte," says Jerry Hickey, a pharmacist and host of InVite Radio, a nutritionally oriented program aired on the New York City-based talk station WOR. "He's sort of a medical detective. He looks at things out of the box, and he's solved quite a few cases over the years."

Like the cranky character played by actor Hugh Laurie on the Fox TV series *House MD*, Bolte is a master of the differential diagnosis—identifying patients' ailments based on a careful analysis of their symptoms. At times cases have taken the doctor into the realm of the weird, as when a patient casually informed Bolte that her psychic told her she was suffering from a parasite, specifically *Entamoeba histolytica*. "I didn't know psychics knew that much about parasites," Bolte says, "but I am always willing to keep an open mind. I said, 'That's fine, but why don't we do a stool test?'" The results proved the psychic correct. Bolte, a spiritual man who is also fascinated by Jungian theory and the workings of synchronicity, chalked that one up to the importance of staying receptive to whatever the universe slings in your direction. The patient, of course, was miffed that she paid for lab work for the doctor to learn what she already knew.

Unfortunately, many of the patients Bolte sees are victims of iatrogenic, or doctor-caused, illness. Simply put, they have been misdiagnosed, overmedicated to the point of sickness, or given treatment inappropriate to their conditions. On occasion, this has led to shouting matches with more conventional docs, like the dermatologist colleague who burst into Bolte's office one day and harangued him—in front of another patient—for telling the mom of an acne-ridden teen to stop feeding her child so much junk food. There's no evidence that diet has anything to do with acne, the dermatologist shouted. Bolte begged to differ and cited the literature. "The pharmaceutical industry has trained even doctors to believe that there's a pharmaceutical answer to everything," he says, shrugging.

Bolte's colleague, Robert Scully, a board-certified internist and medical director of two MediCenter Clinics in Long Island, New York, has followed the younger doctor's work for the past fifteen years. Bolte sometimes consults him for advice. Scully says that there's a very good reason why doctors sometimes miss what is right in front of them. "In mainstream medicine there's an expression: 'When you hear hooves, don't look for zebras.' And as a result, zebras are missed all the time. Tom's a zebra hunter."

Back in 1999, Ethel Moore took a liking to this man who gave her so much of his time and asked a seemingly endless stream of questions. I just know you're going to find out what's wrong with me, she told him. I just know it.

Something about the tone of her voice was familiar. If you listened between the words, you could detect a more urgent plea that Bolte knew too well: *Please, Doctor, help me get well. Please don't let me die.*

Call him Dr. House if you want, but this is not the House you know. You see, before he fixed people, Bolte fixed houses. The houses belonged to his mother, an Irish-American woman named Rosemarie Martin, who bought a handful of dilapidated houses by the seashore in Long Island, hoping to rent them. When the boy was only eleven, his family lost five of its dearest loved ones in the span of sixteen months. A beloved cousin, three grandparents, and lastly, the boy's father—all dead from various illnesses. Bolte's father, felled by a melanoma, had been a dapper Long Island attorney, and when his widow ran out of cash, banks foreclosed on the family's two homes. Rosemarie, who until then had played the part of the doting housewife, whisked her son and daughter away to live in the slums of Staten Island with her father, her children's only remaining grandparent. There was talk of sending the kids to foster homes, but Rosemarie vowed to keep the family together, to get a job, make money, and move back to the Island. She sought financial security in real estate.

The routine went like this: Rosemarie and her son fixed up the houses, the tenants trashed them, and the pair fixed them again. From a parade of handymen, the boy learned to plumb pipes, wire switches, hang drywall, lay brick, and do simple carpentry. He learned that no two tradesmen built a house exactly the same way. They all had their little tricks, their favorite tools, favorite circuits, favorite brands of building materials, and unorthodox pet theories. When something in a house didn't work, you fixed it by running down all the possible options where the system could possibly break down. You found the leak and patched it; you uncovered the rotten joist and sistered it back together.

Years later, his hands would take him to the hearts of machines: jalopies, used computers, musical instruments, sound equipment. When he was still in his early twenties, studying medicine in Puerto Rico and moonlighting at a San Juan hospital, his hands coaxed newborns into the world, sixty or more before he ever graduated from medical school.

During his residency at New York Downtown Hospital, he was chosen to be administrative chief resident of his class—a high honor—in his last year. "He was extremely bright," recalls Satish Dhalla, director of general internal medicine at Downtown Hospital and a professor at New York University School of Medicine renowned for his seminars preparing residents to ace their medical board exams. Concurs Bruce Logan, chief of medicine during Bolte's tenure, "Tom was an extremely reliable, hard-working guy who really cared about the patients. Just a really good guy."

After Downtown, Bolte worked as many as six medical jobs at a time. Doctors, he thought, were no different than builders, and you should try to learn from as many as possible. Inspired by his parents' fascination with nutrition and alternative medicine, he apprenticed himself to Robert Atkins, the famous diet doctor, and Leo Galland, another author of best-selling books who blended alternative with conventional therapies. Bolte didn't think much of Atkins's famous diet at first, but eventually came to think it worked well for patients with type 2 diabetes. Clearly, he realized, med school had taught him a lot about drugs, but his training had been markedly deficient in the areas of diet, nutrition, and exercise. "There's not one drug that has been proven to 'extend' life," he's fond of saying. "The only things that do that are fruits and vegetables."

The first time Rosemarie saw her son in a white coat, a stethoscope draped around his neck, she wept. Somehow, all the time he was in medical school, she had not connected the dots, had not imagined that this kid in gypsum-spattered jeans spackling drywall was going to be a doctor.

Nowadays he remembers how he and Rosemarie crouched atop those houses, the blue of the Atlantic at their backs, nailing shingles to the roof. He made his hands do her bidding. The mother did not know, and the boy did not tell her, that at night in his bed he bargained with God. He had attended five funerals in little more than a year, and they had terrified him. Over the graves of his loved ones he learned the words of the Lord's Prayer for the first time. At night, he prayed: *Please, God, don't let my mom die. Please don't take her from me.*

His prayers were answered. She lived long and prospered. When she died four years ago at the age of sixty-nine, she was a wealthy woman. When she took sick with lung cancer, he gave her the greatest gift he could. He shut down his practice and cared for her 24/7 for the last seven months of her life. "It was the most rewarding thing I've ever done," he says.

THERE IS AN ART TO INTERVIEWING PATIENTS. THE doctor asks numerous questions of the patient, like, "You say you have a throat pain. Is it a stabbing pain or a dull, sore pain?" Depending on the patient's responses and lab evidence, the doctor eventually works his or her way down to the correct diagnosis. In medical textbooks, the diagnostic path is sometimes represented by a tidy decision tree, but in real life the process is rarely so pat. Interviewing patients is laborious work. While you are seemingly carrying on a simple conversation, your mind is sifting through everything you have ever learned about medicine.

More often than not, Bolte is guided by intuition. He tries to bond with the patient as quickly as possible, hoping to forge an alliance in which the patient feels comfortable confiding even the most intimate details. He chatted amiably with Ethel Moore, cracked jokes, told her about himself, his practice, his patients—and gradually disarmed her. Over the course of that first meeting, he learned that she smoked Nat Sherman cigarettes, which she ordered by mail; that she sometimes had a burning sensation on her tongue; that she'd had a history of dental work; that she enjoyed eating fish. A couple of times a week, she experienced a tingling or numbness in her hands. Just before the rash started, she remembered being bitten by an insect at the base of her neck.

Huh, the doctor thought, maybe a biological infection?

He asked her if she spent a lot of time outdoors. She said yes, of course, she lived in the country. She estimated that she'd used hundreds of cans of DEET insect repellent over the last thirty years.

Hmm, the doctor thought.

On and on the talk went, the agreed-upon hour stretching, as it often does, into two hours. When he had looked through her case file, he noticed that her physicians had ordered three biopsies on her lesions, which led to the cancer diagnosis. But they hadn't analyzed her blood for the presence of unusual chemicals. Such a test is hardly standard. It has to be specially requested. On her second visit, Bolte took a blood sample himself and sent it for testing to a lab in Texas that he trusted implicitly. Her previous doctors probably had not requested such a test because she didn't mention anything to them about her generous use of household chemicals. She hadn't mentioned it because they didn't ask.

It's all about the questions—and the time it takes to ask them.

Over the years, to help streamline the Q&A process, Bolte has formulated a questionnaire, which he asks patients to fill out after their first visit. The thirty-two-page document is comprehensive, asking after family medical histories, social history, habits, hobbies, employment, exposure to household products, industrial chemicals, foreign travel, and so on. Bolte estimates it takes two to four hours to complete. He does not apologize for such complicated homework. Humans are complex, and a doctor never knows until much later if a patient's response was significant or just a red herring.

He is currently working with software developers to create a questionnaire program that may help target areas for follow-up by physicians who do not have the luxury of time in which to ask questions—which is to say, nearly all physicians. The economics of modern medicine is not conducive to leisurely interviews. In order to handle all the patients funneled his way by an insurance company, a doctor is obliged to hire numerous staff to handle the paperwork and make sure his invoices are paid. The more staffers on the payroll, the more patients a doctor must see to cover it and ensure a profit for himself and his partners. The more patients he can crank out in an hour, the more profitable he will be.

When Bolte worked in such a practice, he was horrified to find that the HMO actually generated "batting averages"—the average number of patients seen per hour by each physician—which the office manager posted in a back office. Bolte almost always finished dead last.

"There was a time in this country," he says, "when doctors made house calls and when they were done, they sat down to dinner with the family. That's how they got to know you. The next time someone in your family took sick, the doctor knew their context. He knew what their home life was like, knew what they did for a living, and so on. Today, you're lucky if your doctor sees you for twelve minutes. How can you possibly find out all you need to know about a patient in twelve minutes?"

After the HMO experience, Bolte longed to open a practice that was relatively free from the influence of insurance and pharmaceutical companies. Today he literally runs his own office. No secretaries. No partners. No insurance forms. (Many patients later submit his invoice to their insurance companies for reimbursement.) If you call the number listed on his website, it rings his cell phone. "Doctor's office," he says. "What's the problem you're having?" Then, if you agree to his terms—no insurance, $125 for a twenty-five-minute urgent care visit, $290 for a major sixty-minute consult, he'll enter your appointment into his Palm Treo. When you arrive at the office a few blocks south of Bloomingdale's, he answers the door himself, unless he's with a patient. It's like this all day, all week. When it comes time to pay, he runs your debit or credit card through the machine under his desk, prints out an invoice, and asks, "Want me to staple that?"

Although unusual investigations are his specialty, urgent care cases help pay the rent. Hotel concierges near his office know to call him if an out-of-town guest takes sick. He's the official on-call doctor for MTV Networks, and if someone there calls, he skates over to the studios on his Rollerblades with a knapsack of medical supplies strapped to his back. A number of celebrities—an Oscar-winning actress, the widow of a famous rock star—

regard him as their official New York physician. But you don't have to be famous to get him to your bedside. He's one of the last Manhattan physicians who makes house calls. He unhesitatingly gives his number to every patient who visits, even if he knows it could later wreak havoc with his social life, dinners out with his fiancée, or an ongoing home renovation project back on the Island. "Please," he reminded a recent flu patient who was leaving, "I'm completely at your disposal. If you get worse or it doesn't clear up, call me at three a.m. I don't care."

He goes to such lengths to stress this because so many patients have been trained by the medical industry to behave deferentially, as if their doctors are always too busy to bother with them. Yet some of his most fascinating cases were solved precisely because he gave patients the time that their primary care physician could not.

"The first time I met him I thought he was a quack," says Dustin Palmer*, a twenty-nine-year-old sales representative. "I hate saying that because I feel I owe my life to him." Palmer's teens had been hell, punctuated by crippling abdominal pain and alternating bouts of diarrhea and vomiting. Doctors diagnosed him with irritable bowel syndrome. ("A lazy diagnosis," quips Bolte.) Palmer lost so many days of school that it took him five years to graduate from college. By then, he had lost his faith in doctors and was depressed. "I couldn't see any future for myself," he recalls. Dragged to Bolte's office by his mother, Palmer—then a kid with unkempt dreadlocks—rocked back in his chair and uttered profanities at Bolte, who just chuckled.

"Whaddaya laughing at?" Palmer demanded.

"I'm laughing because I *know* I'm going to get you well," said Bolte.

Oh, the kid thought, you met me an hour ago and you know what's wrong with me? Sure you do, buddy.

Bolte's rambling interrogatory style threw the kid off. "I think he's a genius," says Palmer, "but kind of like an idiot savant. He sits there and he's nonchalant, but he's got a lot more knowledge up there than I first expected. You would not expect it." For Bolte, the key was hearing specifically what foods Palmer had eaten prior to his attacks. Over time Bolte demonstrated that the young man was suffering from celiac disease, a genetic disorder that meant he could not digest gluten, a protein found in most grains and commercial food products. As many as two million Americans have the disease, which attacks the intestines, but doctors routinely misdiagnose it. Over the course of several visits, Bolte studied Palmer's food diaries and helped him nail down which foods were most likely to trigger the painful autoimmune response.

Another patient, an art gallery owner, arrived one day complaining of chronic headaches that he'd had for forty years. His primary care physician had him on six different medications, some prescribed for migraines, others for depression, and still others for the side effects from the migraine meds. The suffering was so severe that the patient took to bed for days at a time, tracking his headaches on a calendar, hoping to divine a pattern to the pain. Some days the meds worked, some days they didn't. But no pattern emerged. Every doctor he saw seemed to be grasping at straws: One prescribed an antifungal medication; another, allergy shots. Bizarrely, Bolte seized upon a single symptom: The man casually mentioned an intolerance to egg yolks. If true, this seemingly innocuous point could narrow his problem down to only one or two possible diagnoses, one of them being heavy metal poisoning. Sifting through the file, Bolte found a hair analysis that the desperate patient had commissioned at an alternative pharmacy: The results showed elevated levels of mercury. Yes, the patient said, he had shared that result with his original doctor, who pooh-poohed the test and told him that if he was concerned about it, he might want to reduce his fish intake. Bolte ordered a DMSA challenge test, in which a drug is taken orally that extracts heavy metals—if they exist—from tissues and excretes them in urine. The urinalysis revealed extremely high levels of mercury. The man responded well to medication and a new diet that flushed much of the dangerous toxin out of his body. Two years later, he celebrated his first headache-free month.

Another time, a woman presented with headaches, fatigue,

hives, rash, and countless chemical sensitivities. After a lifetime of fine health, she broke out with the most annoying physical symptoms from using soaps, shampoos, and cleansers.

After a blood workup revealed the presence of hexane, a petroleum derivative known to cause nerve damage, Bolte asked her about her living conditions. What, he asked, makes your home different from other people's? She boasted that she had a fine new home. After living in a one-bedroom apartment for years, she bought and renovated the apartment next door after the birth of her first child. Now she had a roomy three-bedroom apartment, the kind of place New Yorkers would kill for.

What happened to the extra kitchen? Bolte asked her.

Now it's my master bedroom, she replied.

You have a gas stove, right? Where did the old stove in the old kitchen sit, exactly? Can you tell me? Do you remember?

The patient's eyes widened. *My bed is right up against the spot*, she said.

A few days later, at Bolte's insistence, the woman had the gas company inspect her apartment. They located and corrected a leak in the woman's bedroom, not far from her headboard. The patient bought and installed a small sauna unit, which she used to sweat out the toxin. Her symptoms resolved in six months.

Ethel Moore's was the grandmother of all these cases. When the chemical analysis came back from the lab, Bolte was stunned. He picked up the phone. He told Moore that her blood test had uncovered high levels of two toxic chemicals—2-methylpentane and 3-methylpentane—both derived from petroleum products.

Could there be something in her environment that was exposing her to abnormally high levels of petroleum products? Moore said she didn't know how that was possible. She'd already had the local board of health to her home to test her tap water and check for natural gas leaks.

He pressed on. Okay, so we're looking for something outside the house. Have other neighbors been ill? She knew that one man who lived behind her had kidney cancer; that was all. Did she or anyone in her family work in a gas station? Was there a gas station near her home? She said no; she lived in a very nice trailer park in upstate New York, not far from I-87, the New York State Thruway. There was nothing on the site but homes just like hers, all arranged in a subdivision with Hawaiian-sounding names.

In all his years of working on houses, the doctor had never visited a trailer park. He asked Moore to describe what they looked like. How were the homes constructed? In the industry, he learned later, the correct term is "manufactured home." The structure is built off-site and designed to fit on a trailer bed towed by a big-rig truck. Once at the site, the home is craned off the flatbed and placed onto concrete piers or cinder blocks, or permanently mounted to a custom-built foundation.

Which does your home have? Bolte asked. A foundation? Or does it sit right above the ground?

Above the ground, came Moore's answer.

Do you know what was on the site previously?

She didn't. But she promised to look into it and get back to him. He hung up reluctantly, half wondering if he should drive up north to help her, have a look around, and maybe photograph the site for his files. If she had lived there a long time in the presence of contaminated soil, her senses may have become inured to a faint petroleum smell that he would notice immediately.

When she was younger, perhaps, her immune system had been able to fend off the daily onslaught of toxins. But as she got older, they had gained the upper hand. If she spent time outdoors working, living, or gardening, she would have inhaled toxic fumes and consumed particles as part of her daily routine—and at night as she slept.

A few weeks later, Moore called with news. At her urging, the local board of health had pulled the real estate records and discovered she and some of her neighbors were living on the site of a former industrial train depot used by Standard Oil in the late nineteenth century. An independent lab tested the soil for petrochemicals.

Shortly after, the owners of the property voluntarily offered to excavate, remove, and replace the soil under the homes of Moore and some of her neighbors. Moore's lesions had begun to disappear. By January 2000, the levels of toxins in her body had dropped to almost nothing, and she was living a normal life once again.

The doctor wishes he knew more about her, but she dropped out of sight sometime after their last meeting in August 2003. Beyond that, the case file is silent on the Adventure of the Petroleum-Poisoned Senior. Bolte thinks he may have a Christmas card from her somewhere in his files. "That's the problem with my practice," he says. "They get well and I never hear from them again."

Names marked with an asterisk have been changed to protect the patient's privacy. As of this writing, Dr. Bolte continues his practice and house-call service in midtown Manhattan.

YES, SNAKES CAN FLY!

1999

This article was commissioned by a major science magazine, but never ran because, at the time, no one could figure out how to photograph a flying snake. But hey—just because you can't shoot it doesn't mean it's not so.

Twenty-five feet above a grassy field on the outskirts of Chicago, a two-foot-long snake creeps out to the edge of a tree branch attached to a scaffold and flings itself off with all the grace of an Olympic high diver. Down, down, down falls the snake. But then, something happens: in midair it pulls its body up into a distinctive S-curve and undulates furiously, seemingly swimming through air. In this way, it actually gains distance until it sails softly to the ground.

Three video cameras will later confirm what the human participants in this experiment already know: they've got a freaking flying serpent on their hands.

Below the scaffold, Brian Sotio, a thirteen-year-old who adores snakes, scrambles to snag this morning's specimen, *Chrysopelea pelias*, by its tail and stuff it into a pillowcase. Up on the scaffold, Jake Socha, a ponytailed doctoral student at the

University of Chicago who has spent much of the last two years studying the wondrous anatomy of genus Chrysopelea, is nonetheless impressed. "Did you see that?" he asks his pal and cameraman, Eric Duneman. "How cool was that? She really knows how to handle herself in the air!"

Indeed she does, but is this flight?

Sure.

Socha's field, biomechanics, recognizes two different types: flapping flight, accomplished by the beating wings of animals such as birds, bees, butterflies, bats; and gliding, or parachuting flight. This last is the realm of animals that exploit their fins, skin folds, or ribs to turn a likely thud into a deft descent. Every school kid has seen pictures of flying squirrels with furry membranes extended between their fore and hind legs. Equally well known are flying fish, ichthyoid Icaruses that leap above the brine and soar the lengths of football fields to escape predators. But there are even freakier pilots out there, such as flying lemurs and flying marsupials. One of the marsupials, the sugar glider, recently has been embraced in the US as a trendy pet.

But it gets even weirder. In 1996, when Socha was casting about for grad schools, he visited Robert Dudley, a biomechanics professor studying insect and hummingbird flight at University of Texas-Austin, who regaled him with tales of flying frogs with oversized toe webbing that extended their hops. Then Dudley's student, Jim McGuire, showed him a species of lizard he was studying called *Draco volans.* McGuire lifted a specimen out of terrarium and let go of it. The flying lizard extended its cartilaginous, winglike ribs as it dropped a few inches to McGuire's hands. Socha was hooked.

Standing well apart from all these other gliders, Dudley told Socha, were flying snakes. Five species in the genus Chrysopelea, dwelling in the tropical rainforests of Southeast Asia, had evolved a unique way of getting around their arboreal environment. "It struck me as really bizarre," says Socha (pronounced SO-ha). "I think it's the basic reaction. This is an animal that you generally associate with slithering on the ground. These snakes look like regular snakes, they don't have any skin flaps or anything. That's why it's more bizarre."

Granted, the whole idea of a flying snake ranks up there with sea serpents. Ever since that little incident in Eden, humans have assumed—to their great relief—that snakes didn't walk and they sure as hell didn't fly. Yahweh himself gave the order: "Thou art cursed...above every beast of the field," the King James Bible tells us, "upon your belly shall you go, and dust shalt thou eat all the days of thy life." Could it be that some snakes just didn't get the memo?

In the 1890s, scientific literature trafficked in anecdotal evidence that bestowed legendary status upon the snake. Socha uncovered one paper in which a British zoologist recalled how Malaysian natives warned him against tangling with certain "cobras," because "they could 'fly, same like bird.'" (Chrysopelea venom is lethal only to small lizards, not humans.) Another scientist wrote, "They can outclimb the monkey, outswim the fish, outleap the jerboa, and suddenly loosing the coil of their crouching spiral, they can spring into the air and seize the bird upon the wing." As late as 1959, an astonished tea planter in Malabar, India, reported seeing one fly over his head for fifty-five yards.

Aside from a flight study in 1970 by Sukham Pongsapipatana and W. Ronald Heyer, a Smithsonian scientist best known today for his work on declining amphibian populations, and some footage shot by the BBC in 1996, no one has ever seriously plumbed the locomotive abilities of the genus. Socha attributes part of the oversight to the snake's apparent ubiquity. On two trips to Singapore and Thailand to collect data, his project was deemed impractical by numerous government-funded grad students. "'Why study this snake?' they kept asking me," Socha recalls. "'It's so common.'"

On his first trip to Singapore in 1997, he was dissuaded by officials at the Singapore Zoological Gardens from mounting an Indiana Jones-style jungle odyssey in search of the snakes. Don't bother, he was told, they'll turn up. And sure enough, they did.

One sailed out of a tree one morning and startled a jogger in downtown Singapore; a quick-thinking professor remembered the American student's quest and swept up the specimen with dustpan and broom. Another snake dropped into the zoo one afternoon and was nabbed while snacking on a gecko. Five weeks into his ten-week trip, Socha had two snakes.

Those early launches, from the third-story balcony of a building at the National University of Singapore, appeared futile. Socha forced a snake to fly by sliding it out of an elaborately rigged cardboard box. Onlookers reported nothing unusual; each snake appeared to drop to the turf, unharmed, like a stone. Later, a hopeful Socha watched the launch videotapes and saw something astounding: Viewed from the side, the snake did not simply fall straight down, but *glided* to the ground at about a 45-degree angle to the horizon. Neither Socha nor his assistants had ever seen these snakes perform; they expected them to fly like birds. "So you can imagine that from the front it wouldn't look like it had gone too far," he says. "But it was amazing. It was beautiful."

Since then, Socha has presided over numerous snake flights —in Asia and the US—with twelve snakes from three of the five species. Some are better fliers than others, but they all display predictable flight behavior best seen when launched from an ordinary tree branch. Released from its bag, a pretty pelias with red-and-blue markings punctuated by black-and-yellow bands drops two-thirds of its body off its perch in a U-shaped loop, then pauses, its eyes scouting avenues of escape. What's it gonna be: the ground? The scaffold? A distant tree? Finally, head and body spring up, the tail releases the branch, and the snake is airborne.

Within 170 milliseconds of its launch, the snake employs its secret weapon: it flares its ribs and flattens its body from neck to vent, or anus. In one species, *paradisi*, the concavity is quite pronounced; no wonder early explorers likened the snake's flight shape to a piece of split bamboo.

This flattening increases the animal's surface area, but as

Socha hypothesizes, the snake must still do something else to keep from falling like a rock. When any body falls, two forces—gravity and drag—work against each other. Toss a shoe off a building and it catches air even as it falls. The shoe spins, pivoting at its center of gravity, a point somewhere between toe and heel. Likewise, when Socha tossed a non-flying snake out a first-story window, the flabbergasted creature tumbled side over side into a waiting bedsheet below. If a Chrysopelea accidentally drops, on the other hand, it tumbles at first, then pauses in midair, assumes the flight position, and glides away. Socha thinks the snake's lateral undulations—those rapid, midair zigzags—help stabilize it.

One afternoon in Chicago, Socha fast-forwards through a bunch of video sequences that would have startled those early Victorian explorers out of their pith helmets. In one data clip, a paradisi specimen makes a gentle turn—seemingly banking like an airplane. Later, a different specimen tops that by turning 90 degrees in the space of less than a meter. The rotation is effected by tail flip which appeared to send its body into a new direction. "Watch this," Socha says, his finger on the button. Still later, the same snake does a tail flip and turns 135 degrees. "He's rotating," says Socha, his eyes fixed on the screen. "But how does he generate the new force to do that? Momentum is not enough."

Clearly, there are more than enough puzzles to go around. Socha jokes that he's got to answer a handful of them before they let him graduate. For now he concentrates on aerodynamics, the physics behind the anatomy. But at some point he would like to grapple with the bigger, evolutionary issue: How did this happen? "That's the question," says Socha. "How does this locomotor novelty evolve? What are these snakes doing that their non-flying relatives aren't? What do they do differently? Is it that their muscles are different? Does this snake have a different muscle? Or is it just doing something different?"

It's entirely possible, he speculates, that there's no anatomical difference at all. A non-flying relative may well be able to fly—but chooses not to. Says Socha, "People keep asking me, 'Why does it do this?' But maybe a better question is, 'Why don't *other* snakes do this?'"

Jake Socha is today an assistant professor of biomechanics at Virginia Tech. In addition to teaching classes, he continues his research on the gliding flight of snakes.

THE ORGAN MAKERS

2001

Organ donors give the miracle of life to patients in need. But someday, if a donor can't be found, doctors will grow their own miracles in a lab. Meet four brothers who are on the cutting edge of this science.

The doctor they call Chuck stands over the body with an electric blade, ready to make the first incision. The knife whirs, peeling crisp brown skin off the breast and digging into the firm white flesh below. The doctor wields the knife confidently, humming to himself, as if he finds pleasure in severing muscle from bone. His two brothers, Marty the pathologist and Frank the anesthesiologist, stand nearby, ready to offer advice. (Jay, the oldest brother and a pediatric surgeon, is on call.)

It is the last Thursday in November and the "patient" is a twenty-three-pound turkey roasted to feed the assembled Vacanti family. As Chuck buzzes away, slicing white meat and dark, the grown-ups sip Riesling, the kids and Uncle Frank gulp cola. The table is groaning with side dishes. All in all, the scene is about average for Thanksgiving Day.

Still, it does not take long to discover that there is little or nothing average about the Vacanti brothers. Joseph (Jay),

Charles, Martin, and Francis Vacanti work together as researchers in the new field of tissue engineering, a discipline they practically invented. What they are trying to create is nothing less than lab-grown human organs, produced from a patient's own tissue. Their work is urgently needed— roughly 100,000 patients in this country die each year because not enough people donate organs, and many of those who are saved by transplantation ultimately die because donor organs are rejected.

In the new world the Vacantis hope to build, an infant born without intestines and destined to die will get a new gut grown from a clutch of her own cells. As the fruit of the child, that new intestine will never be rejected. Imagine a world, say the Vacantis, in which diseased pancreases, lungs, and spinal cords can be replaced as easily as the transmission in an old Chevy. Imagine a world in which salvation grows in an incubator. Imagine a world in which hope is a given.

So far, artificial skin and cartilage are the only lab-grown tissues available to surgeons. But that is about to change so quickly that even doctors familiar with the research will find it difficult to comprehend the possibilities. It was only in 1996 that Chuck and Jay Vacanti held the first conference of their fledgling Tissue Engineering Society. Today two of their former colleagues, Anthony Atala, a pediatric urologist at Children's Hospital in Boston, and Laura Niklason, a Duke University researcher, have already performed what seem like miracles—Atala successfully implanting lab-grown bladders in beagles and Niklason growing fresh pig arteries in her lab. This year there are more than fifty laboratories in the United States alone racing to create people-made people parts. And the researchers in all of those labs are indebted in some way to the work of the four Vacanti brothers.

HOW DO WE MAKE A SCAFFOLD?

The brothers grew up in Omaha, born to a dentist father and a mother who stopped six credit hours shy of a premed degree to marry. There were eight children, and money was always tight.

Jay, the oldest son, longed to go to Harvard. His father pulled him aside one day and told him that he could work his way through the Ivy League school or go to Creighton, where Dad was on the faculty, for free. Jay reluctantly chose Creighton but soon made it to Harvard as a surgical intern. His brothers got the same speech, also attended Creighton, and eventually followed Jay to Massachusetts.

Meanwhile, working at Massachusetts General Hospital, the largest teaching hospital for Harvard Medical School, Jay witnessed an endless parade of gurneys delivering children who could not be saved. "I have always thought that being a pediatric surgeon was the most gratifying kind of surgery," he says. "You start with the most helpless and vulnerable of humans, diagnose a potentially harmful condition, definitively manage it, oftentimes with surgical intervention, and return the child to his family and to possibly another eighty years of life." But too many children Jay saw needed new livers, bladders, intestines. There weren't enough organs to go around; there wasn't nearly enough hope. Jay Vacanti determined to fix that.

He knew that in 1979 Eugene Bell, an engineer at the Massachusetts Institute of Technology, had grown skin cells in flat sheets of tissue. Jay, along with his colleague Robert Langer, a chemical engineer at MIT, became obsessed with figuring out how to extend Bell's work beyond two dimensions.

The number one challenge in tissue engineering is to get the specimen cells to grow at all. Freshly plucked from a human, a batch of cells hasn't got long to live. They need oxygen, a temperature of about 98 degrees Fahrenheit, and nutrients. So a tissue engineer doesn't waste time. He places the cells in a petri dish with liquid nutrients—carbohydrates and amino acids do nicely —then tucks them away in an incubator. With a little luck they multiply and in a few days produce enough cells to be considered tissue. As amazing as the process is up to this point, it's relatively useless, because clumps of tissue are of little use to a patient. The tissue engineer must convince cells to morph from a meaningless jumble of flesh into a functioning organ.

Jay and Langer's vision was radical—build a scaffold out of

plastic on which the cells could build a three-dimensional organ. "The original organ we had in mind was the liver," Jay recalls. "I thought degradable plastics would make an ideal scaffold. I knew from my work in cell biology that cells adhere to plastic dishes for in vitro culture, secreting their own scaffold as they settle to the bottom of the plate. I also knew you could treat the surfaces of plastics so they would be more likely to cause cell adherence."

But Eugene Bell was already on the record: Plastic polymers simply wouldn't work as a scaffold. He dismissed the idea as *fatuous*. That word stuck in Jay's craw.

In the summer of 1986, Jay took his family to Cape Cod. As his children played in the surf, he perched on a jetty, mulling the problem. You could get cells to populate the exterior of a hunk of polymer, where they had easy access to oxygen and nutrition, but the interior was another story—you might as well try to grow a houseplant inside a basketball. Out of the corner of his eye, Jay saw a sheet of seaweed bobbing in the water. Then it hit him. Nature, the original tissue engineer, had already solved the problem. Under the seaweed's rubbery skin lies a network of fine, hollow branches that pipe fresh oxygen into the organism while pumping out expended gases. Jay shot an entire roll of film of that seaweed and phoned Langer before the day was out. Their scaffold interior had to be light and airy, like spun candy, they decided. Build the branches, and the cells would do the rest.

Today that's the game plan in labs everywhere. Tissue engineers build a porous, three-dimensional polymer model of an organ, squirt a soup of nutrients and living cells over the structure, incubate, and wait. We should add, though, that no one really knows why cells multiply the way they do. This is the big unknown in stem cell research, and something that bears more investigation. "All we have right now is a lot of really interesting phenomenon," another stem cell researcher at Harvard says. "But we don't know the mechanism." Jay speculates "that cells thrive best in this environment because they are designed to live and function in three-dimensional space." In time, the biodegradable framework dissolves in the body's water through the process of hydrolysis, until what is left is wholly alive.

Now when Jay talks to young tissue engineers, he sometimes slaps a transparency on the overhead projector that displays Webster's definition of the word fatuous: "smugly or foolishly stupid." His intention is to encourage his successors to trust their instincts and go forward, ignoring criticism that might hold them back.

THE FIRST HUMAN EXPERIMENTS

In 1989, Chuck and his team, then also based at Mass General, submitted a paper to a top journal announcing they'd grown a piece of human cartilage. It was rejected outright: "No practical implications," the reviewers wrote.

Chuck was stunned. No practical implications? He saw it as a challenge. So he quizzed plastic surgeons: What's the toughest cartilaginous structure to repair? The human ear, they answered, no question. Every day infants are born with underdeveloped ears; children and adults lose ears in car accidents. Somewhere between bone and skin, cartilage is a tricky substance to work with, and the ear is the body's most intricately shaped and visible piece of it.

So Chuck and his team decided to build one. They needed a living host that wasn't human, so they implanted an ear scaffold under the wrinkled skin of a hairless lab mouse they nicknamed *Auriculosaurus*. The mouse proceed to grow an ear on its back. An image of this oddity was beamed into newsrooms all over the world. Facing hard questions from the public and colleagues about his motives, Chuck had to explain that he'd only intended to show the medical world what could be done.

In April 1994, Chuck and Jay got a chance to prove what could be done in a human. They met Sean McCormack, twelve, who had been born with a protruding sternum and no cartilage or bone under the skin of his left torso. Unprotected, his heart could be seen beating just below the skin. As a Little League pitcher, he badly needed a chest wall. Boston's Children's Hospital let the Vacantis conduct a procedure so bizarre the Food and Drug Administration had no regulations to cover it.

The doctors harvested cartilage cells from Sean's sternum to seed a flat, round scaffold about the size of a compact disc. Awash in nutrients, Sean's cells multiplied and permeated the polymer. Weeks later the construct was inserted in Sean's chest. As his body grew, it incorporated the shield for his heart; seven years later, he's a star BMX bicycle racer.

In 1998, Raul Murcia, a machinist, crushed and severed his left thumb in a cargo elevator. Chuck and his team, now at the University of Massachusetts in Worcester, leaped to the challenge: *Think we can grow this guy a new distal phalanx?* Chuck carved a piece of surgical coral into the shape of a thumb, seeded Murcia's bone cells onto it, and in a few weeks he had a thumb digit ready for implanting. By then newspapers were screeching "test-tube thumb," and Chuck was fielding a call from the FDA: *What are you guys doing up there?*

Since Sean McCormack's operation, the FDA had developed a set of regulations governing cultured cells. The surgical coral was FDA approved, and the FDA had decided that one could implant autologous cells—those grown from a patient's own cells—without approval. However, if a doctor combined two FDA-approved technologies, the technique required separate approval. You can implant the new bone, they told Chuck, but you absolutely cannot implant any tissue-engineered cartilage. "I asked what to do, and they suggested I apply for retroactive approval, which I did," Chuck says.

In the end, Murcia got a new thumb-tip, but Chuck was not allowed to attach cartilage or tendons to it. In a paper published in May 2001 in the *New England Journal of Medicine*, Chuck and his colleagues report the case, including the happy results of Murcia's recent biopsy: Much of the coral still exists, but its pores have filled in with Murcia's own cells. More surprising, the new cells are intelligently transforming the coral, remodeling it to look more like a human thumb bone.

Today researchers in tissue labs all over the world have moved beyond thumbs and ears, as they struggle to grow more important and complex tissue structures. In the search for foolproof methods, they play with variables: What's the best polymer for arteries? For tracheae? They jet to symposia, deliver papers, send e-mails to find out what other labs are up to. Such encounters are civil, even genteel, but an undercurrent of rivalry often fills the air.

THE MIRACLE CELLS

The Vacanti brothers have raised sibling rivalry to an art form. Jay knows, for example, that if he's courting an applicant for residency at Mass General, chances are the would-be tissue engineer has already visited Chuck and Marty at the University of Massachusetts, who told him: "If you're offered a position at Harvard, you should consider it just because it's Harvard. If you're offered a position in tissue engineering at UMass, you should accept it because we're better."

Chuck says the brothers are "very competitive" with each other but not destructive. They needle each other in the same delighted way they always have, but they know they are safe. "It's absolutely better to work with your own brother. You trust your own brother in a way you can never trust someone else," says Chuck, who adds, laughing, "If one of us does something to break that trust, we can always go tell our mom."

Their rivalry is balanced by an intimacy few scientists will ever enjoy. "You don't have to go into the dance that you would if you were sharing an idea with a colleague," says Jay, "where both of you are trying to be polite. If you're dealing with your brother, he can say, 'You know, that's really stupid.' It's an efficient way to problem-solve. On the other hand, if he says, 'That's really smart,' you know it's genuine."

That dynamic engendered their most startling breakthrough, which was announced last fall.

In 1996, Chuck had convinced Marty, the pathologist, to leave Nebraska and join him in Worcester. Chuck had grown increasingly frustrated with the fragile adult-tissue cells he had been working with. Most cannot last more than thirty minutes without an oxygen supply. Fetal stem cells are hardier, but harvesting them is controversial.

Chuck told Marty to find an alternative: "Look for stem cells in adult tissue."

He instantly replied: "They don't exist."

"They have to exist," Chuck insisted. "If the human body is constantly trying to repair itself, it must have immature cells somewhere. Find them."

"You're nuts," Marty told him.

"Just do it."

"If I had talked like that to anyone other than a family member," Chuck says, "he would have gone home and told his wife I had an attitude problem." Instead, Marty decided to give it a try. For fifteen months he drew cells from living animals, only to watch them die. He scrounged lab animals other researchers had sacrificed for their work. He scraped flesh with scalpels and dissolved it in enzymes. He peered into the resulting broth, magnified two hundred times, to no avail. At every staff meeting, Marty had nothing to report. It became embarrassing.

Then one day, peering through the microscope, he spotted tiny circular shapes. Adult-tissue cells are about fifteen micrometers wide. Marty saw cells only three micrometers wide. He began showing them around. They're too small to be stem cells, everyone said. Just debris. Junk.

Tired and depressed, Marty stood in his lab staring at flasks of the cell soup, thinking, "Wastebasket or incubator?" For reasons he does not comprehend, he stuck them in the incubator. Three days later, those little specks of junk had multiplied. What's more, they had gone without oxygen for more than an hour before he put them in the incubator, an ordeal adult stem cells could not have survived.

At staff meetings Marty took center stage. Eventually someone asked: What do you call these cells? Privately, Marty had begun to call them "spore-like cells." They had a faintly prokaryotic, spore-like look about them. Until 2.5 billion years ago, life on this planet was limited to bacteria and algae that reproduced through the agency of single-celled bodies called spores, which lay dormant until called upon to create new life. In time, prokaryotes morphed into eukaryotes, multi-celled creatures. Marty's mind reeled when he thought about it: What if the most primitive process of evolution and self-repair was still going on inside our bodies? At one large meeting, on Jay's turf at Mass General, his concentration was broken by Jay chanting sotto voce: "Fungus, fungus, yeast, yeast!"

Weeks later, Chuck phoned with a suggestion, but Marty cut him off. Obsessed now, he had been examining every scrap of tissue he could lay his hands on and had isolated spore-like cells in every one. He'd bought a tray of chicken livers at the grocery. Even there, he found them.

Chuck was agog but couldn't wait to up the ante. Freeze 'em and cook 'em, he said. Marty took the cells down to -121 degrees Fahrenheit. The cells survived. He left them at 187 degrees Fahrenheit for thirty minutes. They lived.

Marty tried to keep a lid on his excitement. He'd learned early that it was prudent to get the data in the bag before you crowed over a new discovery. His confidence soared the day he showed his work to Guido Manjo, an eminent Italian-born pathologist who lectures at UMass. Manjo's advice: Test those cells for DNA—and publish as soon as possible. Then came the ultimate compliment: "Dr. Vacanti," said the senior scientist, "you may have discovered a fundamental process of nature that has not yet been described."

Manjo was correct. DNA was present in the cells, and no one in the history of biology had ever identified such minuscule formations living in mammalian tissue. They were the kind of cells that the Vacantis had been dreaming about: They could live in the body without oxygen for days until blood vessels grew to supply them. Marty's most recent research shows the cells may actually be able to differentiate into tissues other than those of the organs from which they originated.

Properly incubated, they grow like grass on a prairie. The team in the lab at Worcester has used them to grow everything from retinal rods and cones to liver, bone, fascia, skin, and heart tissue. They have pulled spore-like cells out of a diabetic pancreas and grown insulin-producing islets in twelve weeks. They have cut a golf-ball-sized section from a living sheep's lung, stuffed the wound with a scaffold seeded with pulmonary spore-like cells, and watched as the lung incorporated the new tissue in eight weeks. Everyone was in awe: A lung is perhaps the most complex organ in the body, possessing at least half a dozen different types of tissue.

SEEKING TO HEAL THE SPINE

Marty's discovery paved the way for the most astounding experiment at Worcester to date. In the late summer of 1998, the Vacanti team inserted scaffolds seeded with spore-like cells into the severed spinal cords of eight lab rats. They hoped new tissue would bridge the gap. But first, the team cut themselves a big break. Scarring of nerve ends in severed spinal cords interferes with healing. So they put the scaffolds in place immediately after the cords were cut and before scarring set in. The cells quickly stitched themselves into the fibers of the existing cords, and the paralyzed rats regained a significant degree of feeling and movement in their previously paralyzed limbs. "After ten days," Chuck recalls, "you could see little twitches in their toes... In three months, some rats could stand on their hind legs and eat what you fed them."

After several months a few of them walked.

Meanwhile, in Jay's lab at Mass General, Frank, an anesthesiologist and the youngest of the brothers, was building an interest in the same problem. Frank had started out in stroke research, where he made some interesting breakthroughs. He and a colleague showed, for example, that slightly lowering the body temperature of a patient at risk for a stroke, such as during an operation, could protect against irreversible damage and complications. That interest in neurology developed into a fascination with spinal-cord repair.

Frank suspected the severed fibers in a spinal cord wouldn't be able to resist an easy pathway along which they could grow back toward each other. So he used a laser to drill tiny tunnels through his scaffolds, which he implanted into the cords of rats immediately after he severed them.

The procedure fits his personality. Frank loves building things, so much so that he almost didn't become a doctor like his brothers. As a teenager he wanted to be an engineer but noticed that engineers were having a hard time finding work after Nixon decimated the space program. He craved a job where he could work with gadgetry. A medical lab appealed to him, but shifting gears from a technician's mind-set to that of a biologist would be a challenge. Nonetheless, he plowed through academia, taking the Hippocratic oath at twenty-three.

Now a seasoned mechanic, he labors away in the lab on weekends, drilling his 2-millimeter holes in solitude, which he prefers. At first, it looked as if his idea was headed for success. After just six weeks, spinal-cord tissue appeared to have nearly replaced the scaffold. Sadly, the effects on the rats ranged from mixed to negligible. Some of them died; others lived to barely wiggle their toes. "Not enough," says Frank. "I wanna see them jump rope."

Analyzing his failure, he saw that he could improve life-support systems for the recovering rats. Redesigning the study got him thinking about salamanders: "If a salamander damages his spinal cord, he can repair it. They don't scar. But in mammals, the cords form scars... At some point, the ability to scar must have been an evolutionary advantage."

As organisms became more complex, tissues required more oxygen to function. And that, Frank thinks, hurts us in the regeneration department. If a human spine is damaged, the cells cannot tolerate being torn away from blood and oxygen. Scar tissue sweeps in to obscure the damage. If we once had the ability to heal, Frank reasons, we must be able to restore it, and he intends to find out how.

His imagination and curiosity have taken him beyond medicine. He has begun writing physics essays, one of which has been published in a prestigious journal. "You can't advance science unless you take a risk," he says. "It's like fixing a spinal cord. Most people think you're crazy. But to have any success you have to let your mind wander. You have to look for relationships, see how things fit. It's beautiful. But it's not the scientific method."

A HEART OF FOAM

Sooner or later, all tissue engineers are haunted by the body's need for oxygen. It was no accident that the earliest successes in tissue engineering came in hatching skin, bone, and cartilage, which can survive for hours in the body until blood vessels mosey over to attach themselves. "Now we're getting into organs like livers and hearts that are too thick to work that way," Jay says. "To survive, they need oxygen, they need nutrition, they need to dump waste." They need a circulatory system. But how does one coax cells to grow into something as complex as a network of capillaries?

These days Jay, along with researchers at the Massachusetts Institute of Technology and the Draper Laboratory, has embarked on a radical protocol: They are etching blood vessels onto silicon wafers. Cells are then deposited onto the wafers, where they grow into circuits just micrometers in diameter and shaped like branching vessels. In time the entire fragile sheet can be lifted from the wafer and the vessels rolled like cigars or stacked like checkers to build a circulatory structure. They work: Jay and his team enjoy showing visitors a video of blood cells coursing through the man-made vessels like water in a stream. This project has taken them one step closer to meeting their greatest challenge yet.

In a corner of the lab lies a polymer cast of a sheep's heart. To build it, Jay and his team took a real sheep's heart, pumped its vessels full of liquid plastic, then dissolved the tissue in a bath of flesh-eating enzymes. When Jay saw the first cast, it looked like a ball of Styrofoam. "I thought they had made a mistake," he says. "Then we looked at it under a microscope and saw: These are all capillaries. It showed us where we're headed. In organs like the heart, circulation is structure." MIT engineers have used the specs from the cast to design a scaffold, another big step down this road.

Someday, Jay and his team will implant cells on a scaffold and try to grow them into a heart. But that is still a good while off. In the meantime, Jay's lab is working on smaller projects such as building an esophagus for children who are born without one or with a portion missing. Before 1938, this condition, called esophageal atresia, was always fatal. Then doctors began using skin to fashion a replacement. In the 1950s, doctors hit upon the technique used today of replacing the missing part with a section of the colon. However, a high incidence of esophageal cancer is associated with this technique. A grown-to-order organ would be a godsend.

The Vacantis can't rest.

"It was a driving motivation for me to specialize in the surgical care of children," Jay says. "I'm certain that our Sicilian-American cultural tradition imprinted on us the primacy of caring for and nurturing children. We were reared in a family of eight children with a large extended family, including grandparents, great-grandparents, and many cousins."

Growing up in Omaha, the four brothers could not help but pick up that lesson— and more. From their parents, Joanne and Charles, they learned that no theory, no question, no design of theirs was too ridiculous to explore. They grew to manhood knowing in their bones that they did not labor solely for themselves. They've lost their father, who taught them this lesson through a life devoted to his family, students, patients, and science. He died of a heart attack in 1994, and those who gather each year at the table on Thanksgiving miss him. But their mother is still around and does not hesitate to remind them they were put on this Earth to work for the good of others.

Chuck's own heart has led him to an even deeper understanding of the values his parents instilled. Nine years ago chest pains led to an angiogram. It showed that Chuck's left coronary artery had completely occluded. He exercised and shed thirty pounds. Life looked good. Then one day five years ago his heart sounded another alarm as he sat in his office. Despite crushing pain, he calmly instructed his secretary to cancel appointments for six weeks and to call the emergency room while he paged his cardiologist.

"I think I need to come in," Chuck told him.

"I can see you in two or three days."

"Maybe I haven't made clear the severity of the problem," Chuck said. "I believe I'm infarcting as we speak."

He wasn't scared. Denial chased fear from his mind. After another angiogram, doctors told him that three coronary arteries had occluded. Interesting, thought Chuck: My diagnosis was correct. You need a bypass immediately, they said. He stalled, enumerating other procedures. Can you try this or that? Uh, yeah, we can try all those things, but you'll be dead. We're doing this *now*.

So it came down to the healer's own cells rebelling, his sternum sawn and spread, his heart exposed to the glare of lights. Before he went under, he was still curious: I wonder whether I'm going to die.

That's when it hit him. Even in your most desperate hour, it is not just about you. No life ever is.

"When you're gone," he says now, "what you really have is what you're remembered for. I realized I would like people to say, 'He did something good for man. For mankind.'"

ALL FOUR OF THE VACANTI BROTHERS REMAIN ACTIVE in the field of tissue engineering.

ORGANS IN THE SHOP

2005

A few years after the previous story, it was time to sit with one of the Vacantis' collaborators to see how far the tissue-engineering field had come.

Each year desperately ill hospital patients—currently some 89,000 men, women, and children in the United States alone—languish while waiting for a transplant of a liver, kidney, heart, or other essential organs. The doctors who attend them know there are not enough organs to go around. Many patients die waiting.

To improve those odds, tissue engineers are trying to harness the power of stem cells by designing three-dimensional plastic molds, called scaffolds or matrices, that resemble organs or body parts. When a soup of nutrients and stem cells is squirted over a matrix, stem cells may grow into a hunk of tissue that can later be transplanted into a waiting patient. Somehow, the matrix imparts critical organizing information to the cells.

Researchers have successfully created simple tissues such as skin, cartilage, and bone. More complex structures—an ear and teeth—have also been grown. But the hope is that a complex organ, like a kidney or a heart, can be built.

Customized hearts, arteries, or valves would be a boon because substitutes leave much to be desired. Prosthetic devices can't grow with young patients. Donor valves, whether from cadavers or pigs, sentence the recipient to a lifetime of immunosuppressive drugs to combat rejection, and still fail after ten to fifteen years. But a heart, artery, or valve built from a patient's own cells may never be rejected.

In what are among the earliest clinical trials, German researchers have created hybrid replacement valves. They started by removing valves from donor cadavers. Then they stripped the valves of rejection-provoking cells, leaving only an elastin and collagen matrix, and seeded the valves with stem cells taken from a vein in the leg or arm. The stem cells knit themselves into this real-life matrix and, when the valve was transplanted, functioned well for more than three years. Critics of the protocol say it's not ideal, because donor heart valves are still in short supply. So researchers are hammering away on two fronts: nailing down how stem cells work—and constructing better matrices.

Robert Langer, a professor of chemical and biomedical engineering at MIT, is one of the two fathers of tissue engineering. In the early 1980s, he and colleague Joseph Vacanti, a Harvard University medical professor and pediatric surgeon at the Massachusetts General Hospital, were the first to demonstrate a method for growing living tissues by seeding a biodegradable scaffold with human stem cells. In April 2005, Langer won the $500,000 Albany Medical Center Prize in Medicine and Biomedical Research, the nation's top medical award. As Langer explains, designing polymers that work effectively with living cells is harder than it looks.

Q: If I needed tissue-engineered material today, what is on-the-shelf ready for me?

Langer: If you needed skin, you could get that. If you needed cartilage, there are variations of cartilage. But a lot of things are in clinical trials. Among those things in clinical trials are other cartilages, bone, corneas, blood vessels—that's all in humans. Then there's a lot in animal trials: intestines, spinal cords, vocal cords. Most of these use some form of matrix, even the corneas.

Q: What's the main challenge facing you?

L: Well, I wish it was just one! On the material side, it's creating the right materials that are highly biocompatible and that stimulate the right growth and cell behavior. Then there's the reactive side. When you are trying to grow the tissue in a test tube, what are the right media and conditions that you need to use? Do you shake the cells? If so, how? That makes a huge difference. A few years ago Laura Niklason, one of my postdocs, showed that the only way we could make a blood vessel was to shake it. In real life the vessels are hooked up to your heart. So we actually made little tubes, put the different cells on them, and hooked them to a pulsatile pump, which was like a little heart. It pumped, and the valves we made developed the strength to withstand real blood pressure.

Q: So cells respond when they're stimulated in the right way?

L: That's right. A more recent study on heart muscle cells is another example. Gordana Vunjak, who works in our lab, got heart muscle cells to grow, begin pumping, and work much better by stimulating them electrically.

Q: If these work well, why spend so much time designing different matrices? Why reinvent the wheel?

L: For some of the tissues, we want a really elastic matrix. In 2002, Yadong Wang, another of my postdocs, created something called biorubber. It made a really, really rubbery matrix to simulate elastic tissues like blood vessels, heart muscle, ureters, and more. Before that we were using different polymers that were far more stiff.

Q: How do these polymers break down in the body and not cause a problem?

L: We deliberately design them to do that. We build in bonds that will ultimately be hydrolyzed by water. The building blocks —the components of these materials—are generally things already in the body or have been found to be safe in the body. One example is glycerol. Over time the cells colonize the shape, or the action of the body will break down the polymers. It is absorbed and doesn't cause a problem.

Q: How many different matrices have you devised?

L: We have actually created a library of materials. The question is, how do you get different cells to differentiate? How do you get the different cell types you want or get them to grow at the rate you want? For some of these cells, nothing has ever been found that works well. Just to take an example: Nobody has figured out how to get human embryo cells to grow without using mouse feeder-cell layers to stimulate growth. So human stem cells grown in vitro are contaminated. That's just to point out that we're hardly done. There are many unanswered questions.

Q: It's no coincidence that the tissue-engineered products on the market are things that have few blood vessels. The thicker the organ, the harder it is to make because you must build so many blood vessels. How do you hope to build more complex organs?

L: Jay Vacanti and I have been working on that with the Draper Laboratory. We've been working on microfabrication, almost nanofabrication, to create networks of vessels. For some of them, we are etching the vessels on a silicon chip. We're still early, but we've been able to get it to work reasonably well.

Q: How would you get the thickness of larger organs like livers or kidneys?

L: Well, you would grow one layer on a chip or matrix and then stack one on top of another. Hopefully, that way you would get the blood flow over several layers, just like in real life.

Q: How does the US government's limits on stem cells affect you?

L: We use adult stem cells, and we use the human embryonic cell lines that the government has already approved. My own feeling is it's a little bit hard to know how much it curtails you because it's something we can't do. I personally think it's good to go down as many roads as possible to solve a problem, so to have some shut off to me is a bad thing. But we've been able to do quite a bit with the cell lines that are available.

THE LADY IN THE SUITCASE

2002

Scientists continue to debate some of the biggest questions in human evolution: Where did human beings come from, and when? How did they get to Australia? And of course, the biggie: Did Homo sapiens ever get it on with Neanderthals? One iconoclast thinks he knows...

She came to him in 1968, inside a small, cheap suitcase—her burned and shattered bones embedded in six blocks of calcified sand. The field researchers who dug her up in a parched no-man's-land in southeastern Australia suspected that she was tens of thousands of years old. *He* was twenty-eight. Almost every day for the next six months, he painstakingly freed her remains from the sand with a dental drill, prizing out more than six hundred bone chips, each no larger than a thumbnail. He washed them carefully with acetic acid, sealed them with a preservative, and pieced them together into a recognizable skeleton. Looking closely at skull fragments, bits of arm bone, and a hint of pelvis, he became convinced that two things were true. First, the bones were human, *Homo sapiens* for sure, and they once had held together a young woman. As he assembled this "monster three-dimensional jigsaw puzzle," Alan Thorne, then a lecturer in the department of anatomy at the University of Sydney, began asking himself whose bones they might actually have been. He had no idea that many years later, the answer to that question would rock the world of anthropology.

Something else about this woman became clear early on—she had been important and powerful. The pattern of burn marks on her bones showed that after she died, her family burned the corpse, then smashed the bones. Later, they added more fuel and burned the bones a second time. This was an unusual ritual. Ancient Aboriginal women were typically buried without fuss. Thorne wondered if her descendants had tried to ensure that she did not return to haunt them; similar cremation rituals are still practiced by some Aboriginal groups today. As hours and days and months passed, he found himself thinking of her as a living, breathing person who'd spent her life encamped on the shores of Lake Mungo, in New South Wales. If this Mungo Lady turned out to be as ancient as field researchers thought, she would be the oldest human fossil ever found in Australia. To Thorne she was already the most mysterious.

In 1968, most anthropologists thought they had a grip on human evolution: Big-browed, thick-skulled humanoids had descended from walking apes. These hulking creatures were eventually replaced by the more advanced, fine-boned humans of our species—Homo sapiens. Throughout Australia, anthropologists had found only big-browed, thick-skulled fossils. That made Mungo Lady a puzzle. Lab analysis of her remains suggested she was 25,000 years old—old enough to be a grandmother to those specimens—but her skull bones were as delicate as an emu's eggshell. Thorne began to realize that she might be telling him a different story than the one he'd read in textbooks—that the delicate, fine-boned people had reached Australia before the big-brows.

That was an exotic thought, and now, many years later, it is fueling the debate within anthropology over a single huge question: Where did Homo sapiens come from? Most researchers accept a theory referred to as "out of Africa." It holds that numerous species of hominids—beginning with *Homo erectus*—

began migrating out of Africa almost two million years ago and evolved into several species. Then a new species called Homo sapiens evolved in Africa and migrated between 100,000 and 120,000 years ago to Europe, Asia, and Australia, consigning all the earlier hominids it encountered to extinction.

Thorne preaches a subtly different but revolutionary view called "regional continuity." He believes that the species his opponents insist on calling Homo erectus was in fact Homo sapiens, and that they migrated out of Africa almost two million years ago and dispersed throughout Europe and Asia. As he sees it, there was no later migration and replacement: "Only one species of human has ever left Africa, and that is us."

Why does this matter? Because if Thorne and his camp are right, much of what we think we know about human evolution is wrong. In the world according to Thorne, the human family tree is not divided into discrete species such as Homo erectus, *Homo antecessor, Homo heidelbergensis,* and *Homo neanderthalensis.* They are *all* Homo sapiens. Yes, Thorne agrees, from the outside all these hominids looked different from each other, but so do humans today—a Korean, a Nigerian, and a Dane hardly resemble each other. Our ancestors displayed great variety, but they were similar in the only way that mattered: They were the same species, which meant they could have sex with each other and produce fertile offspring. In doing so, they ultimately gave birth to us all.

Mungo Lady started Thorne down the road to regional continuity. Six years after he reassembled her, Thorne and three assistants unearthed another small-boned skeleton only 1,600 feet from where she had been found. At burial, this body had been laid on its right side, knees bent, arms tucked between its legs. Certain features—the skull, the shape of the pelvis, and the length of the long bones—told Thorne he was looking at Mungo Man, which thrilled him. As a general rule, female skeletons are more delicate than male ones, so doubts about the uniqueness of Mungo Lady's delicate bones would be quashed by having an equally delicate male counterpart to study.

Thorne's colleagues took their best guess at this specimen's age, as they had with Mungo Lady in 1968, based on radiocarbon dating and analysis of stratigraphy. They dated him to 30,000 years ago. As the oldest humans ever found down under, the finds were considered so important that the Australian government declared the sandy, bone-dry crater that was once Lake Mungo a national park in order to honor—and protect—the site. To the Aboriginal tribes, the pair became precious symbols of their early peopling of the continent.

But Thorne assigned a meaning to the bones that resonated beyond Australia. To his mind, the presence of two such unusual skeletons suggested that the peopling of the Pacific was a richer, more complex process than anyone had ever imagined. Anthropologists had long assumed that the first Homo sapiens to reach Australia were fishermen who left Indonesia and got blown off course, ending up on the new continent. Thorne began to wonder whether the first journey from Indonesia to Australia was not an accident but an adventure, undertaken with confidence by intelligent, mobile people. Mungo Lady and Mungo Man closely resembled skeletons of people living in China at the same time. Had these people migrated in boats to Australia? Had there been successive waves of immigration by different peoples over tens of thousands of years? To imagine such things, Thorne had to abandon familiar notions of early man as a blundering primitive.

He had already begun to do so. In the months he'd spent piecing together those braincases, he had begun to think of them as his elders, worthy of respect, capable of thought and imagination. That supposition was not an outrageous one for an Australian anthropologist to make. From childhood Thorne had grown up on a continent that was home to one of Earth's oldest continuous cultures. He'd learned a great deal about Aboriginal culture while working his way through college as a reporter at the *Sydney Morning Herald.* From where he stood, the ways of Mungo Man and Mungo Lady were not so different from those of modern Aborigines. He could easily picture two different tribes settling near Lake Mungo, one from nearby Java, another perhaps with roots in China. And once the two parties were encamped around the lake, it was not hard to imagine them crossbreeding.

Those who believe in regional continuity tend to have a view of human sexuality that is more generous and more inclusive than that of the out-of-Africa proponents. In the latter view, Homo sapiens led a kind of search-and-replace mission as they spread around the planet; these researchers believe that members of the new species would not have been able to successfully reproduce with members of earlier species, no matter how hard they tried. Thorne thinks that's nonsense. "European scientists have dominated this field for 150 years," he says. "And they've got a big problem in Europe. Namely, they've got to account for those Neanderthals. My opponents would say that Cro-Magnons"—humans identical to us who lived during the Ice Age —"simply 'replaced' Neanderthals with no intermingling. That's the part I object to. 'No intermingling.' Now, I ask you, does that sound like the human beings you know?"

In the early 1970s, these ideas were pure speculation. Thorne had no proof of anything. The bones had told him what they could and then lapsed into silence. So he tucked them away and went on with his career. Three decades later, the bones spoke again.

In 1997, Thorne finally got the tool he needed to explore Mungo Lady and Mungo Man further. European scientists reported that they had successfully extracted fragments of mitochondrial DNA (mtDNA) from the remains of Neanderthal skeletons unearthed in Germany, Croatia, and Russia. This was stunning science; the Neanderthals had died out 35,000 years ago, and yet researchers had been able to harvest genetic matter from their bones as if they'd expired yesterday.

It was the beginning of a revolution in paleoanthropology. Geneticists were hooking up with bone men everywhere. They were focusing on mtDNA because the mitochondria, which lie outside the nucleus, are easier to study—in a human cell there are only thirty-seven mitochondrial genes compared with 100,000 genes found in the nucleus—and because it is the only DNA anyone has been able to isolate and interpret in ancient fossils.

For reasons not yet understood, mtDNA survives the ravages of time better than nuclear DNA. And it has another interesting attribute: It's inherited only through the maternal line. Scientists have seized upon this characteristic to try to build genetic family trees. Almost two years ago, geneticists working in Sweden and Germany reported studying the mtDNA of fifty-three living people from around the world. Within this small sample, they found that Africans shared a characteristic sequence of mtDNA, and that everyone else carried at least some portion of that sequence in their cells. The research suggests that all living humans had their roots in Africa. But Thorne doesn't put much stock in this report. He thinks the conclusions are questionable because samples taken in Africa today could be from people whose ancestors were not African.

When the first Neanderthal studies were published in 1997, Thorne had already retired. He had traveled the world for thirty years, excavating sites and filming science documentaries for Australian television. His face and his ideas were as well known in Australia as Carl Sagan's once were in the United States. At the request of the Aboriginal council, Thorne still safeguarded the Mungo fossils. Because three more-sophisticated dating technologies were now available, he ordered new tests on thirteen of the individuals in his care, and the results gave him a shock.

The ages came back first. Using the new technologies, his team found that the small-boned Mungo Lady and Mungo Man were actually 60,000 years old—twice as old as anyone had guessed. Thorne saw these dates as a crushing blow to the out-of-Africa theorists. No matter what his opponents said, there wasn't enough time on their 120,000-year clock for Homo sapiens to leave Africa, dash up to China, evolve from rugged Africans into small-framed Asians, invent boats, sail to Australia, march to the interior, get sick, and die. How much simpler everyone's life would be, he thought, if anthropologists could agree that some of the players in this drama had reached China 1.5 million years ago and continued to evolve there.

After the dating, Gregory Adcock, a doctoral student in genetics at Australian National University, decided to check all thirteen fossils for mtDNA. But first he set up stringent procedures. It's easy to contaminate specimens: More than once, scientists have been embarrassed when the "ancient DNA" they extracted turned out to be their own. To avoid this catastrophe, Adcock alone handled the specimens. He alone traveled between two testing labs. He sampled his own DNA and Thorne's to use as a control. Before sampling the ancient specimens, he tested five modern human and animal bones to make sure he'd perfected handling techniques. Then he drilled into each fossil and took a sample from the bone's interior, where no one could ever have touched it. Of more than sixty samples he analyzed, he reported only three contaminations. Ten of the thirteen individuals yielded DNA.

The results were nothing less than remarkable: Among the ten successful extractions was the world's oldest known human DNA—plucked from none other than Mungo Man. (No DNA was recovered from Mungo Lady, perhaps because she had been cremated.) Mungo Man also appeared to mock the findings of previous scientists: His mtDNA signature did not match anyone's, living or fossil, on Earth. There was no evidence that he was genetically related to ancient Africans.

The findings were published in January 2001 by Adcock, Thorne, and five other researchers. What followed was intense disagreement. "People just fell over when they read this new stuff," says Alan Mann, an anthropologist at Princeton University and a moderate in the human-origins debate. "The people at Mungo were totally modern looking and were expected to carry the DNA we have, but they didn't. I think that makes for an incredibly complicated story. It's a stunning development."

Thorne's critics were underwhelmed. "Alan is great at generating media interest. He's a former journalist, you know," says Chris Stringer, head of human origins at the Natural History Museum in London, a staunch advocate of the out-of-Africa model who is accustomed to his phone ringing off the hook every time Thorne fires another volley. "He has done some important work. I'm not saying his work is bad or wrong or whatever. Obviously, I have a different interpretation of it."

Stringer and his colleagues laid into Thorne. First, they said it was unlikely that ten of the thirteen skeletons had yielded mtDNA. This was an unprecedented success rate, so they believed that there had to be contamination. Even researchers at Oxford University, in one of the world's finest labs, had contaminated specimens. The detractors also pointed out that mtDNA lines died out all the time; the Australians were making much ado about nothing. This part was true: 25 to 30 percent of mankind's mtDNA has been lost over the past million years when women gave birth to boys or didn't reproduce at all.

Thorne concedes that mtDNA has evolved greatly over time, and all scientists working in this area have to be cautious. But as long as everyone is using mtDNA analysis as a basis for speculation, he asks why his work alone is regarded with such suspicion. Mungo Man and his alternative complement of genes were alive enough to make it to Australia and contribute to the peopling of a continent. Modern Aborigines didn't inherit Mungo Man's mtDNA, but they have certainly inherited the characteristics of his skull. "Eventually, all these people intermingled, and that's why the Aborigines have such diversity," he says.

Stringer, for his part, maintains that the out-of-Africa model can indeed account for a settlement in southern Australia 60,000 years ago. Africans, he says, would have had to travel only one mile toward Australia each year for 10,000 years to make that possible. In other words, the Homo sapiens who left Africa 100,000 years ago would have reached Indonesia with plenty of time to sail to Australia.

In New York, Ian Tattersall, one of Thorne's closest friends, has long quibbled with his stance. "We've agreed to disagree," says Tattersall, curator of anthropology at the American Museum of Natural History. "I have a lot of respect for him; I just think he's barking up the wrong tree." Tattersall argues that Neanderthals were so obviously a separate species that Homo sapiens could not have bred with them.

Thorne says his lifelong study of animals has taught him otherwise. In captivity, for example, jaguars have mated with leopards and pumas and produced fertile female offspring—

although all three animals supposedly belong to different species. Polar bears and brown bears, wolves and coyotes, dromedaries and Bactrian camels also cross-mate. Darwin himself dismissed the word "species" as a term that is "arbitrarily given, for the sake of convenience."

In recent months Thorne and his team have examined every human fossil from Australia and Asia they could get their hands on. They're retesting their Mungo Man work, hoping to confirm the findings and fill in some of the remaining gaps in the fossilized man's mtDNA profile. To satisfy their critics, they are allowing three rival labs to analyze Mungo Man extractions. Results will be available by the end of this year. When they are, they will most likely be debated. This science is still too inchoate for either side to declare victory.

Whatever the outcome, the bones from Lake Mungo have created change in Australia. The nation has committed to returning Lake Mungo and its environs to the Aborigines. Soon elders of the tribes living around Lake Mungo will decide when they will assume management of the land, artifacts, wildlife, and tourist trade. In 1991, standing near the metal stake that marks the spot where Mungo Lady was found, Thorne returned her bones to the elders of those tribes. At the time, elders debated whether to rebury her or preserve her. Thorne argued for the latter. "If you do away with her bones," he told them, "I'll always be right. You won't be able to refute my work. Someday there will be an aboriginal Alan Thorne, and he'll have a different way of looking at these bones. You have to give him that chance." The council voted for preservation. Today Mungo Lady inhabits a vault at the Mungo National Park exhibition center that can be opened only when two keys are present. Aboriginal elders hold one; Thorne was presented with the other.

Despite Thorne's proselytizing, only a small fraction of the world's anthropologists accept his theories. But he couldn't care less. These days, he draws inspiration from the old Sherlock Holmes maxim: "Once you eliminate the impossible, whatever remains, no matter how improbable, must be the truth."

He points out that regional continuity is by far the simpler theory and can much more comfortably account for all the complicated twists and turns in the genetic evidence of human evolution now coming to light. "It argues that what is going on today is what has been going on for two million years, that the processes we see today are what have been going on in human populations for a very long time. You don't need a new species that has to extinguish all the other populations in the world. This is why out-of-Africa is the impossible, and regional continuity is not only not improbable but the answer and the truth."

In 2003, scientists appeared to accept new tests, which showed that Mungo Man and Lady were about 40,000 years old. And it's now believed that humans lived at the Mungo site between 50,000 to 46,000 years ago. Though more scientists now say that Homo sapiens and Neanderthals interbred, the debate is far from over. Sadly, Dr. Thorne died in 2012 at the age of seventy-three.

THE SCIENTIST WHO THREW THE LIGHT SWITCH

2003

Hey—who turned on the lights? A young scientist rocks the world of biology with an astoundingly simple observation: evolution owes everything to the invention of the eye.

I'm standing in front of a massive painting by Hans Holbein the Younger that shows two sixteenth-century diplomats, posing next to a table crammed with books, scientific equipment, and musical instruments. Painted on the canvas below the regal pair is an odd, elongated shape. Like most everyone else around me, I have no clue what I'm looking at. Andrew Parker is squatting on his haunches near the bottom corner of the painting, scuttling like a crab across the gallery floor. He tilts his head, looking at the painting from an oblique angle well outside the frame. He motions for me to join him, and when I do, the two men in the picture fade from view, and the long shape morphs into a huge grinning skull. "That's so cool," is all I can say. Parker just smiles and says, "Mmmm." He makes this thoughtful sound about a hundred times a day.

We've been in the National Gallery in London for a couple of hours. It's my first time but Parker, a zoologist, knows the place like the back of his hand and treats me to a running commentary on the permanent collection. The day before, as we walked around Oxford, where Parker teaches, he told me that he read the entire Sherlock Holmes canon as a child. That makes sense. Hanging out with Parker, I feel like Dr. Watson trailing lamely after Holmes; the bumbling, average Joe who sees the world the way everyone else does, following a cool-headed logician with the soul of an artist. At the scene of the crime, Watson observed only what was there; Holmes always saw what wasn't, and what should be. Parker is like that: his eyes see what others don't. He knows to look at the Holbein from below, nearly on his knees. Sherlock followed clues to nab felons. Parker chased color and light, and ended up plugging holes in Darwin's theory of evolution.

The man with kaleidoscope eyes has managed to crack a mystery that evolutionists have fretted over since Darwin first sharpened his quill: What triggered biology's Big Bang? Half a billion years ago, jellyfish, sponges, and worms bobbed on the waves or slogged blind through the muck of endless seas. But in virtually no time those pulsating blobs of protoplasm blossomed into all the animal groups, many with spectacular hard parts such as shells, spines, antennae, and joints. This is called the Cambrian Explosion, and it's the sort of instantaneous flourishing of life that drives scientists mad.

The way the Watsons see it, you just don't go from four phyla to thirty-eight, or develop internal body plans, in an evolutionary wink. It doesn't seem possible. Darwin pored over every fossil he could in an effort to understand the mystery, but the Cambrian remains mocked him with their diversity. And when he turned to the Pre-Cambrian fossils that had lived only five million years earlier, his eyes beheld the same three or four chunks of slug meat etched in stone. It was infuriating. "The case at present must remain inexplicable," he sputtered. More than a hundred years later, Stephen J. Gould concurred: "How in heaven's name could such a disparity have arisen in the first place?"

Enter Andrew Parker, age twenty-nine. In 1996, the young Brit—then living in Australia—introduced what he calls the Light Switch Theory: The Cambrian Explosion was triggered by the appearance of the very first eye. With vision, he argued, came a new dynamic between predator and prey, between male and female, and with that, a whole new set of strategies to survive. Change was key: change of shape, size, behavior, and first and foremost, color. Overnight, the scales fell from the eyes of every Tom, Dick, and Watson in the biological sciences. Parker's theory seemed so...elementary—and yet no one had thought of it before. A few years later, the Royal Society would sweep in and set Parker up as a lecturer at Oxford University, and the *London Times* would dub him one of the three most important young scientists in the world. This spring his new book *In the Blink of an Eye* might very well make him a scientific celebrity.

I meet Parker on a Friday in Oxford. Walking through the city center and onto the campus, I feel as though I've entered a landscape of gray and grayers. The architecture is imposing and muted, with an old-world appeal. I imagine Parker in one of the university's fabled, ornate buildings with gothic spires and stained glass windows. But when I find him, he's stuck in an ugly cinderblock structure with evident asbestos issues. We have a few hours before Parker's due to lunch at his college. "Why don't we have a coffee in the canteen?" he says, and soon he's leading me up stairs, around corners, through corridors of white cinderblock in the zoology building.

There had been a tentative plan to meet at Parker's house on the outside of town, but it didn't pan out. He'd lived right in Oxford until a few years ago but the endless parade of meetings, lunches, and dinners proved too much. "I'd end up staying out all night and getting no work done," he says. Finally he bought an old Victorian house that he could work on himself. When I ask him about it, he doesn't describe anything but the color—which he chose—and the light. The lounge is apricot with French doors opening north to the garden, bringing in natural light. The kitchen is yellow, and there's an image of a large lemon tree that he painted on the west wall. An old farmhouse table stands on beige tile, worn smooth with time. The main bedroom is lime green and full of old wooden furniture. (I'm told his girlfriend approves.) The bathroom fixtures gleam white amid dark blue tiles. The study where he wrote his book is Tuscan yellow and sports a blue carpet. There are few overhead lights, and a scattering of floor and table lamps illuminate the rooms. I take a chance and ask him what his favorite color is. "Well, it depends..." he begins.

That's what I thought...

At the canteen we grab a coffee and some shortbread biscuits, and I notice that the only ornament in the cafeteria is an old portrait of Darwin, who regards me staidly, hat in hand. The theory of evolution was first debated on this campus. The Bishop of Oxford himself came out to rail against the new idea.

We sit down and Parker, now thirty-five, tries to give me an idea what his Light Switch Theory is all about. "All the laws of life that we have today that are governed by light, all come down to the fact that eyes exist," he says. I wonder if I'm missing something in the glaring simplicity of this statement. But he goes on. "Without eyes, those laws would just completely vanish," he says. "So that very first eye must have suddenly written all the laws. You only need one eye in the environment for everything to have to adapt to light. You don't have to have eyes yourself to adapt to light, you just need eyes to be there."

He's talking about the laws of survival: *Stay alive! Eat, and avoid being eaten!* When the lights came on, creatures could see each other for the very first time, and everyone looked like somebody's lunch. They had to adapt or die. One of the best ways to stump a predator is to change your color, and that's the strategy many living things have adopted ever since. Parker learned this firsthand when he was twelve. His parents took him and his younger brother to Greece, where they could run on the beach and soak up the sun. Parker loved to snorkel, to watch fish and other marine animals darting around the sea floor. So there he was, splashing around in warm Aegean waters, when he saw a yellow fish stuck in some rocks. He reached down to rescue the fish—and the freaking rock moved! It was in fact a moray eel, a species that has evolved to blend into the sandy sea bottom.

"The animals didn't know it themselves," Parker continues. "They had no idea what was going on, but they were evolving towards light. Even if you don't have eyes yourself, if you're fluorescent pink, you'll just be picked out by predators and disappear. So for the survival of the species you have to become adapted to light. But that situation didn't exist before the first eye. Before the first eye there was no such thing as camouflage and adaptation."

Wait, I ask him, why couldn't I live as a worm in the Cambrian? Our earliest ancestors, I've read, were worms with spines. And they were blind.

"They burrowed into the sediment and hid completely," he says.

And if they didn't do that?

"Then you're a chunk of protein. You're a meal, basically."

Parker gets a sly grin on his face as he trots out this ghoulish line. The world he sees is equally blessed with horror and beauty. He's been thinking about this law-of-the-jungle, kill-or-be-killed-type stuff since he was a boy growing up in Birmingham. His great-grandfather was an amateur naturalist who lived on a small farm in Kent, the countryside southeast of London. He had a family library in the house that went back two hundred years. Whenever the Parkers went down to visit, Parker chucked aside Holmes, Dracula, and Frankenstein to flip through musty volumes on biology. The old man used to take him out for walks, looking for badgers and foxes. Parker collected freaky insects that looked like walking sticks and took them home as pets.

"I had different stages, really," he says. "Early on it was natural history, then I was more into art, then into music. My father taught me guitar. He loved the guitar. Rock and roll, mostly. The Beatles. The Stones." All through college he played lead guitar with Head in the Heavens, a band from Birmingham. They were doing well, and preparing to tour with UB40 when Parker bailed at the last minute. "It was really strange," he recalls. "I liked it for a long time, but one day I just had to quickly stop, as soon as possible. I just didn't want to play in front of all these people anymore." Parker says he bores easily, a trait he regards as an irritating part of his character. To this day he struggles with this tendency to bounce from project to project. Right now he's forcing himself to finish the scientific papers he's got floating around in various stages of completion. I ask him how many papers, exactly: at least twenty. Most scientists are lucky if they've got three balls in the air at a time.

When he was ten, Parker got a set of oil paints for his birthday and started squeezing pigment onto palettes. "At school I was told to get into art or design as a career," he says. "I think I got the highest marks in art the school ever had. I always painted. A lot of nature scenes. Outdoor scenes. Landscapes. I want to get into it again now. I just bought a new set of paints."

He learned early on that there's more to painting than just daubing pigment on canvas. There are numerous tricks you can use to produce an effect. A skillful artist, for example, can play two or three colors next to each other, or layer them in a way so the viewer sees a color that isn't quite there. Parker saw this trick in nature, which endows such creatures as beetles, butterflies and seashells with what Parker calls "structural color."

We head back over to his office so he can show me what he means.

He shows me these gorgeous green and blue beetles encased in hard plastic. If I tilt them, the color changes, as if in a hologram. He keeps an iridescent blue butterfly, *Morpho didius*, at his fingertips that does the same thing. The pigment in the Morpho and the beetles is really black, brown, gray, and other dull colors. But the screaming neon I see is an optical trick, light waves bouncing off sophisticated structures in their shells and wings.

It seems funny to be talking about color and light. The powers-that-be have consigned Parker to a room with a large window that looks out over a concrete wall. He makes up for the view with the dozens of postcards of Australia lined up on a ledge in his office. A couple of stills of the actor Peter Sellers sit next to the postcards. Classical music plays softly in the background. "Mmmm," Parkers says absently, as he starts reaching for specimens on his desk.

For Parker, looking is an emotional act. Potential seems to lurk in any image he encounters. I'm surprised by the presence of the fluorescent lights overhead. Parker says he tolerates it. Some nights when he comes home from work, he'll put on a little music, pour himself a glass of wine, and move through his house adjusting the lights. His girlfriend has come to accept this as part of his unwinding ritual. Even if he's just dropping off his stuff before heading to the gym or the soccer field, he's still going to be messing with the lights. Sometimes, when he fantasizes about leaving science, he thinks about becoming an artist who sculpts with light.

Parker isn't the first biologist to study structural color in animals. Long before he arrived on the scene, scientists had identified a couple of mechanisms that animals used to produce color. They knew, for example, that a beetle's shell behaves like a crystal, reflecting only certain colors of light. Parker's genius lay in finding another color mechanism, and another, and another. His first find was while doing marine biology work in Australia. He was looking at ostracods, minuscule pale crustaceans, under a microscope. "I was just moving them about," he says, "and I noticed this flash of light. A green light—and it was so bright. I thought there must be some explanation for it. I called over a few people to take a look because I wanted to make sure I wasn't seeing things."

He wasn't. In his early college days he was scrambling to settle on a major and had studied optics in a physics class. He learned that you could split light into a rainbow of colors by etching microscopic grooves on surfaces with sophisticated instruments. The grooves are diffraction gratings, and they're so difficult to make that they're employed as anti-counterfeiting devices—in credit card holograms, media copy-guard stickers, and the colorful sheen on the surface of CDs. No one had ever described diffraction gratings on the body of any living animal. The idea that a bottom-sucking scavenger was manipulating light in a way that humans thought they invented in a lab was a revelation, not to mention a blow to the collective human ego.

Parker chased that color with a microscope until he found diffraction gratings in the animal's antennae. That alone was a big discovery, but the day he spotted a couple of ostracods pairing off in a bucket, it was all over. "These things had never been seen mating before, and I thought, gosh."

Gosh is not the word I'd use if I saw two centimeter-long crustaceans humping in a bucket, or ever really. But it seems to suit Parker. What he witnessed was a mating ritual: the male would flash his colorful antennae to signal the female. When Parker saw that, he knew color was hugely important to the survival of the species. "When these things have a function," he says, "it means that they are a product of evolution." Think peacock tail feathers—or neon hot pants.

Soon, he was finding gratings in other animals, and even ancient fossils. *Cambrian* fossils. He mounted a cast of these fossil gratings on a piece of glass, and submerged them in seawater under sunlight. They flashed brilliantly. He had lifted color from the canvas of time. "For the first time we could reconstruct an extinct animal in color," he says. "That was quite amazing, but in a way it was just sort of an interesting..."—he pauses, searching for the right word—"fact."

He was already bored, in other words. He says that when he was painting, he used to start one picture, get about halfway through, and start something entirely different. The only reason he made the leap to science and marine biology was that he thought they would discipline his mind. He looked to his father and uncle, who were both engineers. If he could just turn to the hard stuff, he thought, to math, or to science, he might be able to chase the boredom away. "I was wrong about that," he admits now. "I got to Australia and was doing things I shouldn't and going off in all directions." I wonder if Parker would be where he is today if he hadn't been off in all directions.

As he was typing up the paper about his discovery of Cambrian color, his boredom egged him on. Who cares? he thought. Was this a truly important find—or was it scientifically trivial? Was it, as he likes to say, more useful to Spielberg than to science?

He had to step back from the canvas and look at the problem again. If a tree falls in a forest, ears are required to hear it. Well, doesn't color make a similar demand? Does it not require an eye?

If evolution gave ancient animals color, it was because someone or something was there to *see* it. "And then I realized—that very first eye which evolved must have had amazing consequences for life," he says. "Evolution is driven in lots of cases by light. Everything is colored. Everything is camouflaged. Everything is adapted to light, wherever you see it. If you look into fields, there are thousands of animals there but you can't see anything because they're all adapted to light. If they're not adapted to light, they'll be eaten quickly and they'll become extinct."

Parker's theory offers a convincing explanation for all the different kinds of animals that appeared in the Cambrian explosion. He says that it's not necessarily a case of going from four to thirty-eight phyla. The different body plans weren't born then; the internal blueprints for all living things were there all along, hidden inside soft-bodied creatures. The only thing that changed was the armor, color, and behavior these creatures adopted to make themselves less visible—and less palatable.

When he first started talking about the Light Switch Theory, it achieved the rare outcome of delighting both physicists and biologists. Physicists appreciated the methods Parker used to arrive at his theory. They loved the idea that these astounding little animals had evolved optical systems that put the best human systems to shame. Biologists looked forward to incorporating the news into what they knew about animal vision. Paleontologists, the same people who had been studying these fossils for a century but had completely missed the boat, thought he was over-emphasizing vision. How important could eyes be, really? they stammered. Now many evolutionists accept the theory as the best one they have. Today the most critical thing anyone has to say about the Light Switch Theory is that it's painfully obvious. When the idea was about to debut in the *Sydney Morning Herald,* an irate editor called the reporter who had interviewed Parker. "Are you sure no one's ever said this before?" the editor demanded. The writer called Parker, who just laughed. He was getting used to that reaction.

"It *is* so simple," he says. "The reason no one's noticed it before is because the fields of paleontology and vision research have been kept completely separate, with a lot of other subjects between them. All you have to do is put the two together, and you get the answer."

In other words, all you had to do was look.

And now I'm doing just that. We head off for lunch and I'm looking at everything. The veil of gray seems to have lifted: Oxford is a sea of pale yellow sandstone buildings; the sky a watercolor blue behind purplish, charcoal clouds; the trees silver and denuded; the grass damp green. We cut through a park and St. Giles Church, where lichened gravestones tilt in the muddy umber sod. At Parker's college, we suck down limp fish and chips with the profs while undergrads sit at long wooden tables. Parker's the youngest professor I see. He nods politely as the woman next to him drones on about John Locke. He stares at his plate and murmurs, "Mmmm." The cavernous room is lined in dull brown wood. Somber portraits of past principals grace the walls. Gray January light spills over Parker's shoulders. He's by far the most colorful object in the room. I imagine that if I were going to paint a portrait of him, I'd have to find an art store that sells tubes of good English oils, and I'd stock up on the blues. Cerulean blue for his shirt, Winsor blue for the snug crew sweater he's wearing, and a rich cobalt blue for his eyes. I watch him choose from the ridiculous array of cutlery, and think I'd mix pale colors for his skin and hair, and save the pinks and whites for his nails, which are meticulously trimmed to the curve of each of his fingertips.

Parker introduces me to some of his colleagues—the history lecturer, the French teacher, the bearded guy who teaches human sciences. While we eat, Parker tells me about the paintings on the walls around us. Until the early 1990s, Somerville was a women's college. Margaret Thatcher went here. So did Indira Gandhi. A little while later, we polish off our bread pudding with cream ahead of the others, and look around. Down the table from us, a guy is eating a plum with a knife and fork. Parker grins and stands up: "Shall we?" I'm right behind him and we head out.

It's pouring. Parker looks unruffled, his gray-blue windbreaker zipped tidily to the neck. I tell him the lunchtime ritual reminds me of a Harry Potter movie, and he lights up. He hasn't read the books but he can hardly miss the phenomenon. "That was filmed just down the block," he says, "I thought we'd walk around a bit. I'll show you."

Oxford is an intellectual theme park. He shows me the famous pub where J.R.R. Tolkien and C.S. Lewis had more than their fair share of pints, points out the building where evolution was first debated, and finally sneaks me up to the great hall featured in the Potter movie. We stand there watching as hungry undergrads file in, dressed in jeans and disheveled black robes. Dishes and cutlery clatter away. The dark paneled walls are covered with old paintings. Parker says that when he first came to Oxford, he couldn't help but notice the grand old buildings but now he's gotten used to them. Architecture was never his thing.

We dash through the rain to a tiny, school-run art gallery. I ask Parker if England's miserable weather or its muted palette had anything to do with his obsession with color. He's staring at a painting of Henry VIII. He calmly rotates his head toward me and feigns a look of horror: "What gives you that idea?" I try to explain: His is a nation of naturalists and gardeners, all drawn to color, many of whom have enacted an age-old drama of leaving home, gathering specimens, and returning to make sense of it all. For Darwin, it was the Galapagos. For Burton and Speke, it was Africa and the source of the Nile. For Parker—

"Australia," he says later, as we pick our way over slippery cobblestones. "Mmmm. Beautiful place." We would see countless works of art in London the next day but this is the only time I would hear him use the word beautiful.

When he was twenty-two, Parker finished college, ditched his bandmates, and set off to travel the world. He got as far as Australia. Stepping off the plane, feeling the warmth on his back, he noticed immediately that this was a far different place than he'd ever been. "It was a place of opportunities, not just for zoologists but for everyone," he says later. As soon as he could, he threw himself in the water. "Everybody talks about the Great Barrier Reef and you have this impression that if you want to see marine life in Australia you have to go the Great Barrier Reef.

Even the people who live in Australia head there. But it's everywhere." Parker was in his element. Animal and marine life, heartbreaking color and endless light were at every turn. He couldn't leave. His world tour would have to wait. He would eventually spend ten years there, land a job at the Australian Museum in Sydney and apply for dual citizenship.

The next day in London it's almost sunny. I meet Parker in front of the National Gallery and we head inside, and the first painting I stop at is a Venetian painting of the Grand Canal. He comes over and takes a look. "It's nice, but the water's not very good. Wait till you see Turner." He's referring to J.M.W. Turner, the nineteenth-century artist, famous for his landscapes and imperiled ships. "I love Turner, and not because he's English. I'll show you."

We find several Turners in one of the east wing galleries, filling the room with sky, water, cloud, and steam. I stop at one that shows an old battleship being towed out of the sea for the last time. The ship is already a ghost. The sun is setting in an orgy of color, and casting the sea aflame. The strokes of the sky are laid on thickly, almost with a knife. Parker looks like he's just slipped into a hot bath. "He had this ability to dab it on and make it look like something," he says, waving an imaginary brush in the air.

He leads me around to the Impressionists and we look at them all: Monet. Manet. Seurat. Pissarro. Renoir. Degas. He's especially taken with Pissarro. It's the strokes that grab him, tiny dabs of color that seem to be arranged on the canvas artlessly. With your nose pressed to the canvas, you see nothing but a meaningless jumble of color. But when you step back, the colors rush together seamlessly into a recognizable image. That's the way it goes, Parker says, in television screens, newspaper photos, and in biology. Later that afternoon, we take the Tube to London's Natural History Museum, where he shows me a model of a common housefly's compound eye. Each facet of a fly's eye sees only one pixel, rendered as one color, at a time. Flies can detect the slightest movement because they're living in a pixilated world. If a pixel changes, danger's coming and they're out of there. Parker looks at the model with admiration. "Flies turn everything into a Pissarro painting," he says.

As Parker bounds ahead of me up the stairs, I think about the way he sees things—his kaleidoscope eyes, the orchestra of lamps in his house—and I think it could only have been Parker who restored color and light to our understanding of evolutionary theory. The day before, he told me he's planning more books about color in nature and the Cambrian. He hinted too that he's got yet another evolutionary theory up his sleeve, but he doesn't want to talk about it.

From where I stand on the ground floor, I can see him darting around a catwalk above me, inspecting some aircraft dangling from the ceiling. Watching him is like trying to track one of his shimmering butterflies.

We end the day at a Mexican restaurant, of all places, in Covent Garden. Parker needs to get to a party at a friend's house. And I need to make my way back to Oxford tonight. We say good-bye, and he disappears into the crowd. I pass up the shiny black cabs lined up at the curb and smile as I step onto a brilliant, red double-decker bus.

Today Parker is research leader of the zoology department of the National History Museum, London, and a visiting research fellow at Green Templeton College, Oxford. Besides In the Blink of an Eye *(2003), he is author of* Seven Deadly Colours *(2005) and* The Genesis Engima *(2009).*

THE WEATHER INSIDE YOU

2000

Is the weather making you crazy? Some scientists think they know why. Non-believers think they're the crazy ones.

John S. Westefeld has seen lots of strange neuroses in his career as a counseling psychologist, but few strike him as odd as those exhibited by patients who become agitated and anxious as soon as they hear predictions of storms. As the bad weather gets closer, they get more and more upset, calling in sick to work, hunkering down indoors with their eyes glued to the Weather Channel, and even refusing to leave the house to pick up their kids at soccer practice. "They are immobilized by fear," he says. "Even if the danger is way down the line." Westefeld has a name for them: Severe-Weather Phobics.

Maybe we should have seen this coming. After all, everyone knows rainy days make us blue and sunny days make glad our hearts. And an increasing number of scientists are saying peoples' reactions to weather are more than psychological. Researchers called biometeorologists and bioclimatologists believe shifts in temperature, humidity, cloud cover, wind speed, and barometric pressure have direct, precise effects on health and behavior, and they believe our bodies are hardwired to pick up cues from the weather.

It's a concept that spans millennia. Even Hippocrates suggested that weather was responsible for patients' ills. "South winds induce dullness of hearing," he wrote, "but if the north winds prevail, coughs, infections...occur." In his folksy book *Weathering*, physicist Stephen Rosen rattles off a list of other famous believers in the mind-body-weather phenomenon, including Columbus, Darwin, Ben Franklin, Johannes Kepler, Blaise Pascal, and Leonardo da Vinci. All attributed their own behavior and mood shifts to what was happening in the sky. Writers and poets, too, have long limned a connection. In a moody Philip Marlowe short story, Raymond Chandler wrote about miserable Santa Ana winds whipping down the mountains and making everyone in Los Angeles think of murder. "On nights like that every booze party ends in a fight," he wrote. "Meek little wives feel the edge of the carving knife and study their husbands' necks. Anything can happen."

Such weather-behavior linkups wouldn't be much news to populations in Pacific Rim nations, the Middle East, and Europe. Germans are especially big believers. For more than a decade, Deutscher Wetterdienst, the German national weather service, has issued daily *bio-wetter* bulletins to the general public on the order of "cloudy with a chance of migraines and damp with a chance of insomnia." Similar bulletins are headed for your local weather forecast. The Weather Channel already broadcasts a daily "aches and pains index" in winter months. Weather Services International, the world's largest weather-news gatherer, has debuted a popular biometeorology page on its website, Intellicast.com. And at least three other major meteorological companies, such as Accu-Weather, are considering their own foray into weather-behavior forecasts.

Can that be surprising in a nation in which more than half of all ill people use alternative medicine and herbal remedies? "Back in the 1980s, health stores were considered part of the lunatic fringe," says Joe D'Aleo, the meteorologist who launched Intellicast's weather-behavior page. "Now we run into our neighbors there. Americans are interested in anything that might promote better health."

Still, the claims of biometeorology are difficult to confirm. Although medical conditions from angina and arthritis to hip fractures and dental periostitis have been statistically linked to weather, statistics alone are not proof of cause and effect. "I'd rank this stuff up there with the signs of the zodiac and old astrology texts," says Dr. John Renner, a family physician and president of the National Council for Reliable Health Information. "Put it this way: Would they be willing to ground a pilot or a plane based on the current evidence? Let's separate the hunches and impressions from the hard medical facts."

Surprisingly, many biometeorologists agree. "We have—and I exult in this—the Missouri attitude: Show me!" says Dennis Driscoll, a biometeorologist at Texas A&M University. "I would like to see both doctors and meteorologists take the bit in their mouths, so to speak, and start doing some definitive research."

But how? Designing such experiments is practically impossible. "I had one patient tell me, 'The frost is on the pumpkin and the pain is in my joints.' How would you test that?" says Terence Starz, a clinical professor of medicine at the University of Pittsburgh. "It would be too invasive. You'd be sticking measuring devices in peoples' joints."

A classic belief—found in folklore and in old biometeorology texts—is that there is some connection between falling atmospheric pressure and the onset of labor in pregnant women. Bad weather hits, and out pops your kid. Is it true—and can you test it scientifically?

In a 1997 study, doctors at the University of Massachusetts Medical School compared maternity ward admissions data from the Worcester, Massachusetts, station of the National Weather Service. Their conclusion: "Although there was an observed statistically significant association between falling barometric pressure and the onset of labor, the magnitude of the difference is not of clinical significance."

"Hurricanes, too, have been associated with the rupture of the amniotic membrane," says Sharon Phelan, an obstetrician at the University of Alabama at Birmingham, who, like many doctors, has been present at births that occurred during bad weather, "but it's all anecdotal. A study with real patients? I don't know how you would do that."

Well, there is the work of Michael Persinger.

LOCKED IN A STEEL CHAMBER IN THE BASEMENT OF A campus building in Sudbury, Ontario, a student sinks into a ratty brown chair, snuggles up with a blanket, and pulls on a yellow motorbike helmet studded with wires. Outside, Persinger, an American-born neuroscientist and clinical psychologist, sits at a computer screen directing a series of electromagnetic waves at his subject's temporal lobes. With a thousand tests already in the bag, he has assembled a repertoire of patterns that predictably induce moods volunteers later describe in questionnaires as "mild euphoria," "low level fear or panic," "depression," or "drowsiness."

"We're not sadistic here," Persinger says, "so most of the time it's pleasure."

Persinger places his subjects in an acoustic chamber the size of a bank vault to block out other electromagnetic waves. The pulses he pipes in are 1 microtesla, about as powerful as the magnetic field you'd get sitting 10 centimeters away from a computer screen. Persinger believes not only that the pulses mimic subtle electrical signals carried by pre-storm air masses, but also that such electromagnetic fields are forces the brain reacts to. "What tells a cell what to do?" he says. "What affects chemicals and behavior? Electromagnetic patterns."

At Laurentian University, where the thin, three-piece-suit-clad Persinger has taught for thirty years, students joke about the prof's chamber of horrors, and he himself can't resist references to Dr. Frankenstein. Still, he is arguably North America's preeminent biometeorologist. To him, it makes perfect sense that living things should be susceptible to weather-related stimuli. Researchers already concur that the absence of sunlight can trigger seasonal depression. And many accept the word of elderly patients who complain that wind, with its shifting sound waves, irritates them so much they'd rather stay indoors. But most physicians aren't likely to agree that gases released from soil during turbulent weather could affect their patients' moods. Or that Earth's magnetic fields could alter body chemistry. To Persinger, they're not just conceivable, they're probably measurable.

In his behavioral neuroscience lab, students bombard rats with extremely low frequency (ELF) electromagnetic waves, which Persinger believes mimic electrical charges carried in weather systems. Critics say such waves are too low in intensity to have any effect on humans. But Persinger insists it's not the waves' intensity that matters but their frequency and shape. Sooner or later, the body's autonomic nervous system, the bundles of nerves and fibers that control our bodies' internal workings, will respond to the faint but persistent barrage.

The mind is constantly adjusting to tune into the information around it. Consider what happens at a typical cocktail party, says Persinger. When you enter a room filled with people talking, the voices sound loud at first, then as your hearing acclimates, the voices seem to mute. The human mind also learns to sit up and pay attention to low voices as well. "If I'm talking to you and I raise my voice and RAISE IT AND RAISE IT, it gets so loud that you leave," says Persinger. "If I whisper, though, you pay attention and you can understand what I'm saying." Whispers slice through background noise because they carry two pieces of information: the metamessage ("This is important") and the actual message ("Do this"). He believes extremely low frequency waves grab the brain's attention in much the same manner.

Day and night, in tiny cinder-block rooms in Persinger's animal lab, furry, pink-nosed subjects scurry around Plexiglas cages wrapped with coils of blue wire like large electromagnets. In one experiment, rats were induced to attack their neighbors after exposure to extra low frequency waves. In another, rats with multiple sclerosis found their symptoms suppressed after electromagnetic bombardment. In still another experiment, rats first taught the location of food and exits in a maze were then exposed to magnetic fields for an hour. Placed back in the maze they couldn't recall any of their new skills. The results dovetail with biometeorological lore; for years adherents have claimed that weather events, such as approaching storms, can enhance or disrupt learning.

In a 1997 study, Rod O'Connor, one of Persinger's students, found that the incidence of sudden infant death syndrome seems to be higher when geomagnetic storm activity is very low, with background frequencies of about 0.2 to 0.5 hertz. (Geomagnetic storms are caused by the compression and decompression of Earth's static field by solar winds.) Researchers have found that infants who die of SIDS have low levels of melatonin, a hormone that is best known for controlling our internal clocks but also mediates production of nitric oxide, a chemical transmitter integral to breathing. O'Connor believes that magnetic fields generated during weather events can depress nocturnal levels of melatonin and in turn, nitric oxide, disrupting infant breathing. To test his hypothesis, O'Connor has been exposing rat pups to weak magnetic fields of no greater than 0.5 hertz, similar to those that occur during geomagnetic storms. Pups exposed to the lowest intensity fields die soon after showing brain chemistry changes that reflect a drop in melatonin levels.

This is robust, interesting science, but aside from Japanese and Norwegian public health officials requesting information, O'Connor's findings haven't generated much attention from other scientists. And that leaves the twenty-eight-year-old neuroscientist frustrated. "The big focus in sudden infant death syndrome right now seems to be on placing infants on their backs as a way to reduce the chance of death," he says. "That's a practical solution, but it doesn't explain the underlying mechanism of the trouble. The attitude is: 'Why look for anything else?'"

As Persinger and O'Connor well know, biometeorologists learn early to put a lid on their own exasperation. History has not been kind to those who indict weather.

In the early 1930s, Chicago pathologist William F. Petersen

blamed the weather for changes in the blood clotting and white blood cell count of three patients. In 1943, Petersen was insisting that weather was destiny. The greatest men were those conceived during when sunspots flared and bad weather ran rampant. Of course Abe Lincoln beat Stephen Douglas in the 1860 election. Lincoln was conceived in miserable weather; sunny skies reigned as Douglas was conceived. The rail-splitter had spent his whole life building an immunity to adversity, while Douglas's corpulence had shielded him from life's hard knocks. At the time, Petersen's theories ranked coverage in *Time* magazine. His work has since been forgotten.

A more typical fate befell Joseph Hollander's weather studies. In the 1960s, the rheumatologist placed arthritic patients in a climate chamber to see if simulating bad weather and increasing humidity would cause them pain; it did. Hollander included his findings in a textbook he edited. Rheumatologists still refer to the classic text, but the weather chapter has been neatly excised.

And Persinger, who believes that electromagnetic pulses are the mechanism behind everything from religious visions and ghostly sightings, has been denied grants so often that he subsidizes his lab mostly out of his own pocket. He arguably is a bigger hit among TV producers seeking footage of his quirky experiments than he is with other researchers. He jokes that medical acceptance of his theories and methods is like the weather: "It comes and goes."

Other biometeorologists see the issue of acceptance differently. Laurence Kalkstein, a bioclimatologist at the University of Delaware and an expert on heat and human mortality, sticks to the provable: Heat kills. In 1995, with a grant from the Environmental Protection Agency, Kalkstein developed software that told officials in Philadelphia and Washington, DC, when dangerous heat waves were on the horizon and how many people might die. He is doing the same for Shanghai and Rome, with grants from United Nations agencies. "I agree biometeorology has been through an identity crisis," he says. "But now it's gaining recognition."

Biometeorology matters, Persinger says, because the way in

which we order our lives in the twenty-first century leaves us especially vulnerable to weather insults. "Our indoor environments have become extraordinarily physically constant," he says. "The lighting. The temperature of our houses. One thing we do know about biological systems is that they tend to respond to contrast. And one of the things that fluctuates most these days is the weather."

Today, if a huge low-pressure system—one of the last forces humans can't control—comes screaming out of the west and squats over the eastern seaboard, in the two or three days when the mass is parked overhead, thousands of people will fight with their spouses, wreck their cars, give birth, and fail exams. No scientist can study such a sprawling matrix and convincingly pin the rap on weather. But Persinger envisions a day when people answer a daily web-based questionnaire. With such data in hand, biometeorologists might then have a leg to stand on. And maybe we'd know for sure that our faults are not in our stars, but in the clouds.

PATRIOTS ABOARD THE BIG BLUE BUS

2010

These young soldiers fought for their nation on foreign soil. Back home, they are fighting a different battle: convincing America to go green.

Ten years ago, Robin Eckstein was a college student in Appleton, Wisconsin, struggling to pay her bills with bartending and waitressing jobs. Her credit card debt was mounting. Out of the blue, a National Guard recruiter e-mailed her, offering a free college education in exchange for her military service. She enlisted, and reported for active duty in October 2000. Three years later, she was driving supply trucks across the Iraqi desert.

Eckstein, thirty-three, has been out of the US Army for three years now, but on this frigid Tuesday morning in late March, she finds herself pulling transport duty once more, this time driving a big, blue, biodiesel-fueled bus across the state of Ohio, from Columbus to Cincinnati. She is ferrying veterans like herself: Matt Victoriano, an ex-Marine; Rafael Noboa Rivera, a former Army sergeant; and Nick Anderson, a former Army specialist. The foursome is making its way through the Midwest as part of Operation Free, a campaign to promote clean energy organized by a progressive leadership institute called the Truman National Security Project. Operation Free, now in its second year, includes dozens of vets who have logged more than 25,000 miles traveling across twenty-three states, stopping at union halls, factories, statehouses, and radio stations and making appearances on nightly news programs.

At each stop the veterans get off the bus and share their stories, eyewitness accounts of the ways in which America's dependence on oil affects not only which wars the US fights but also its ability to wage war. In their own words, the vets say what many people have said before: America must become energy-independent, invest in renewables, and commit to a future that eradicates the threat of climate change—not because it's the feel-good thing to do but because this nation's security may depend on it.

These vets' views have become increasingly mainstream, even among national defense experts. James Woolsey, director of central intelligence under President Clinton, is outspoken about the connection between the dollars the United States pays to satisfy its oil addiction and the enemy ordnance lobbed at its troops. "Except for our own Civil War, this is the only war that we have fought where we are paying for both sides," Woolsey has said. "We are paying for these terrorists with our SUVs." And in spring 2010, thirty-three retired generals and admirals signed an open letter to the leaders of the Senate, stating that "America's billion-dollar-a-day dependence on oil makes us vulnerable to unstable and unfriendly regimes." They called on President Obama and Congress to "enact strong, comprehensive climate and energy legislation to reduce carbon pollution and lead the world in clean energy technology."

Robin Eckstein is tall and blond, stylish in her black suit. As she drives the bus down I-71, she tells me in her flat Wisconsin accent what a typical day in Iraq was like. As an Army specialist, a notch higher in rank than private first class, she rolled out of camp every morning in a slow-moving convoy of trucks carrying water and fuel to troops dispersed throughout the desert. Exposed to harsh weather and frequent sniper fire, her detail was and is one of the most dangerous in the service. "We were the weakest link," she says. "If one of us gets taken out, you don't know how far the dominoes are going to fall." Without fuel to power up their Humvees, helicopters, and tanks, troops can do little other than sit and wait. Even now, she says, she cannot help but think of thirsty soldiers waiting for fuel in the middle of the desert.

Off the bus, several days and several states later, I meet with Aaron Scheinberg, who tells a similar tale in Cambridge, Massachusetts, not far from the Harvard campus, where the twenty-nine-year-old former Army captain is a graduate student. "There were times in Iraq when we couldn't go anywhere," Scheinberg says. "We couldn't have medical helicopters escort us because we ran out of fuel. Other times soldiers were killed bringing fuel to our base." Just starting a tank's engine—from a cold start to the moment you're ready to roll—consumes seven gallons of fuel. Running at a top speed of about 45 miles an hour, the average Abrams tank gets a paltry six-tenths of a mile to the gallon, he explains.

Scheinberg is the New Jersey coordinator for Operation Free, and when he's not taking classes at the Kennedy School of Government, he's driving to his home state, volunteering his time and energy to speak at events much like those Eckstein and her crew attend. Scheinberg is one of a half dozen veterans on campus who take time away from their coursework in business, government, and international relations to talk with the media and the public about their shared conviction that clean energy goes hand in hand with the United States' ability to maintain its competitive edge in the world.

Drew Sloan, thirty, is in his second year at Harvard Business School, and when we meet up near campus, I notice a faint scar on his right cheek. In 2004, when he was serving in Afghanistan, he was part of a two-vehicle convoy zipping along a riverbed when his Humvee was hit by a rocket-propelled grenade. Sloan woke up four days later in Walter Reed Army Medical Center in Washington, DC, recalling nothing of the attack that blew in his vehicle's windshield and shattered most of the bones in his face.

In October of that year, he was awarded a Purple Heart and a Bronze Star. He then turned down a medical discharge to go to Iraq, where he earned a second Bronze Star.

In late 2009, Operation Free asked Sloan to testify before the Senate Committee on Environment and Public Works. He argued that climate change and its attendant extreme weather events would only exacerbate geopolitical instability. As sea level rises, he said, millions of people who live in coastal nations such as Bangladesh may have to flee across borders and become climate refugees. If that happens, who would quell the unrest? "The US military is the only institution that can take on a massive humanitarian crisis," Sloan tells me.

"Whenever anyone attacks the science of climate change, they ridicule the data as being *uncertain*," Sloan says. "Veterans know you can't wait for 100 percent certainty. If you wait until everything is clear and laid out, you're probably no longer alive. Or if you are still alive, you've definitely lost because someone else has seized the advantage. Veterans know how to deal with ambiguity and still make decisions."

His testimony hews closely to the US military's own recently adopted positions on climate change. In the February 2010 Quadrennial Defense Review (QDR), a summary of Department of Defense strategies and priorities, the department discussed climate change for the first time in the report's history. "Assessments conducted by the intelligence community indicate that climate change could have significant geopolitical impacts around the world," reads the document, "contributing to poverty, environmental degradation, and the further weakening of fragile governments." In an effort to reduce its own carbon emissions, Defense Department officials laid out a series of goals that might surprise most Americans. For example, the Air Force will increase its alternative fuel load to 50 percent by 2016; the Army is converting non-tactical vehicles to hybrids and electrics stateside; and the Navy is exploring biofuels for its carriers.

The last night I'm with the bus in Cincinnati, Noboa, Anderson, and Eckstein are talking to a group of about one hundred people gathered at a labor hall for a green energy and jobs rally. Some people wear hard hats that read, "2 million green energy jobs now!" Others sport T-shirts that say, "Make our energy clean. Make it American."

The vets have been on the road since 6:00 a.m. They held a morning press conference, shook dozens of hands, briefed aides to Ohio governor Ted Strickland, and spent a few harried hours trying to fix a broken brake light along the way. Eckstein is exhausted when she steps up to the podium, but somehow manages to draw on the crowd's energy. She speaks with conviction as she recalls details from the QDR, telling the crowd that the Department of Defense and the Central Intelligence Agency are serious about taking action. "These are not organizations known for hugging polar bears," she says.

The line gets a laugh from the crowd, but a bearded man in his fifties isn't buying it. He finds it laughable to extol the green leadership of the US military. The military is creating a human and environmental disaster in Iraq, he says. He advocates dismantling the armed forces and using that money to rehabilitate the Midwest's factories and invest in projects that promote a "humane" green economy. "The military is not a leader," he says. "The military is the obstacle!"

For a moment, it looks as if the riders of the big, blue bus have crashed headlong into the idealism of the old left. But Eckstein responds respectfully. "Thank you for your passion," she says. "It's not that I don't care about the human toll, because —trust me—I do, and I know my fellow veterans do. But if this is the approach we have to take so that certain other individuals will get it, is it not a good approach? Certain individuals, when they hear the words 'climate change,' they shut down. For whatever reason, when they hear veterans speak on it, they actually listen," she says. "They get it, they understand it, and they're willing to change. That's what we want. We all want change."

She is taken aback by the applause, and blushes as she takes the next question. Later, as the hall empties out, the man who challenged her is among the first to approach Eckstein to shake her hand.

THE TOWN CHANGED BY WIND

2007

For it or against it? A stranger brings wind power to a small town, and everything hits the fan.

In the spring of 1999, a stranger named Bill Moore arrived in the small town of Lowville, the county seat of Lewis County, New York, and checked into the Ridge View Motor Lodge on Route 12. Like many towns in this county, Lowville (the "Low" rhymes with "now") is blanketed with countless parcels of farmland and pastureland that are whipped, especially in winter, by powerful winds gusting out of Canada, sixty miles to the north. This is dairy country, though the constant pressure for cheap milk has made the economics too chancy for single-family dairies. Moore, a broad-shouldered Yale man who had worked on Wall Street, had a passion for renewable energy and owned a company, Atlantic Renewable, that had already developed two wind farms—one with seven turbines and the other with twenty—seventy-five miles south of town. He had what he considered a good proposal for the 27,000 citizens of Lewis County: Milk wind, not cows.

When he started telling the locals about his notion, Moore was met with indulgent smiles but little genuine enthusiasm. "I made a presentation at a town meeting, and they looked at me like I was from Mars," he says. "They were polite. They didn't openly laugh." Next, he wandered the countryside, knocking on farmhouse doors to ask permission to erect meteorological equipment to test the wind speed. "I was thrown out by a number of people," he says. "They've listened to a lot of sales pitches over the years: seed salesmen, fertilizer salesmen." As he'd learned over the years, farmers were a tough crowd to win over.

Eight years later, though, it's as if the cool reception Moore received never happened at all. Windmills stud this flat, stark landscape as far as the eye can see. Each turbine is taller than the Statue of Liberty, and nearly all of them are spinning inexorably toward the future of Lewis County—and perhaps the US's own. This is the Maple Ridge Wind Farm, the nation's largest new alternative energy project east of the Mississippi River. In the last year or so, 195 turbines have become operational in the towns of Lowville, Harrisburg, and Martinsburg, capable of producing 320 megawatts of electricity, the amount generated by a medium-size power plant, or enough power to run 98,000 homes.

What lured Moore and kept him commuting between Lowville (pop. 4,548) and his home in Maryland was the knowledge that this part of Lewis County was uniquely qualified to become the East Coast's largest wind farm. From the back of the motor lodge, he could see the land rising to a kind of rocky ridge. This is Tug Hill, which has stood up to harsh Canadian air skating off Lake Ontario as long as there has been wind, stone, and soil. The plateau, about six hundred feet higher than the surrounding countryside, sits west of the Adirondack Mountains in farm country. The Irish, German, and Polish immigrants who settled here a century ago did so because the land was cheap, and they soon learned why. The plateau is colder than the surrounding lowlands, and the growing season two weeks shorter. A few hundred feet above the plateau, wind speeds of 100 miles per hour are not unknown.

You wouldn't know it from the oversize fiberglass cow that stands optimistically beside the local farm co-op store, but the milking industry here is in decline. There are a few employers in the area—factories that make Kraft Philadelphia-brand cream cheese, bowling pins, gift boxes, and book covers—but they hardly provide work for everyone. The sons and daughters of the dairy farmers are leaving in droves for jobs elsewhere. The scenery has fallen into something resembling an Andrew Wyeth painting: Collapsed silos, dilapidated barns, and abandoned farmhouses are everywhere.

Moore felt strongly that his offer could bring new jobs, stabilize tax revenues, and elevate the quality of local schools, but to make it happen he would have to persuade more than seventy-eight landowners to let him collect his data by erecting giant turbines on their land. He had to help the citizens of Tug Hill fall in love with wind.

Bill Burke, a fifty-nine-year-old dairy farmer, belongs to the fifth generation of Burkes born on Tug Hill. He and his wife, Patricia, live in an Italianate-seeming white house with black shutters that is perched on the edge of the plateau, with wind turbines to the front and back. Burke has raised and milked cows on his 598 acres for more than thirty-six years and was just getting around to formulating his "exit strategy": Sell off two hundred head of cattle, sell off equipment, sell off land, and settle down to a retirement blissfully free of debt. "I don't know if you know what it's like for us farmers," he says, "but most of us are in debt from the moment we start up in business. I bought out three farmers when I got started, and now there's no one left to replace me." He wouldn't go out of his way to encourage his son or daughter to follow in his footsteps: "No money in it."

Maybe, he thought, he'd substitute-teach at the school where his wife was principal. Or maybe he'd sell seed or feed. Then along came Moore, touting wind. "I was on board from day one," Burke says, "but I could see others would take some persuading. We'd hold meetings and no one would show up. Not because they opposed the idea. They just didn't think it would happen." Noting that there are more than a hundred landowners in town, Burke says, "Getting them all to agree on something is like trying to herd ducks."

To educate themselves about Moore's proposition, Burke

and his wife drove south to Madison County to inspect Moore's seven-turbine wind farm; five of the turbines are on the property of dairy farmers Carl and Bonnie Stone. The Burkes parked in the Stones' driveway, as many visitors do, and started asking questions. The Stones, who are unfailingly patient with the tourists, engineers, college students, and wind enthusiasts who come to gawk at their turbines, welcomed the Burkes in after hearing their story. Emboldened by what they learned, the Burkes soon agreed to allow the developer to collect data and ultimately to erect seven turbines on their property. The guaranteed income—a minimum annual payment of about $6,000 per turbine, adjusted annually for inflation—has transformed their lives. "It's paying for me to retire," says Bill Burke. "It's given us a chance to stay in our house," adds Patricia Burke. "We don't have to sell after all. We sold off the herd one spring, and the heifers later, and now we just have to decide what to do with the land." (They plan to lease to area farmers.) For now, Bill has given up the notion of selling seed. He and Patricia travel the state stumping for wind power.

At the heart of their acceptance of such a project in their own backyard was money. They had spent their lives dealing with the economics of the milking business; now their future, and that of their neighbors, was linked to the arcane economics of wind power. The money behind wind power is as capricious as the wind itself. On and off for the last two decades, Congress has dangled subsidies like fluttering kite tails in front of eager wind developers, hence the inherently stop-start nature of the industry. Some years you get it, some years you don't. As a result, some years you build, some years you don't. All major energy producers get subsidies from the federal government, but Big Wind's take is puny. In 2003, the United States doled out somewhere between $37 billion and $64 billion in write-offs, low-interest loans, income and sales tax breaks, construction bonds, and so on to the energy sector. Of that, Big Wind—companies engaged in industrial-scale wind power generation—received perhaps 1 percent. The juiciest wind subsidy—the Production Tax Credit (PTC), which credits wind developers 1.9 cents for every kilowatt-hour they produce—is scheduled to expire at the end of 2008. Congress will most likely renew the PTC, but not before making Big Wind sweat it out. Meanwhile, developers are rushing to get wind farms operational before the deadline so they will be eligible for the credit.

In some states a second important economic tool is making wind power a virtually risk-free investment. In 2004, New York established what is known as a Renewable Portfolio Standard (RPS), a law requiring power utilities to purchase at least 25 percent of their electricity from renewable sources by 2013. The deadline is absolutely critical: Number crunchers estimate that by 2010 New York State will not have enough power to run itself. Critics say the RPS is artificially inflating the green power market, but New York is hardly alone: at least twenty other states have passed similar mandates.

So for the moment, a wind power boom—boom with a lower-case b—is on. Wind power generating capacity in the United States grew by about 25 percent last year, and we'll likely see the same percentage growth this year, if not more. US wind power plants are now capable of producing 31 billion kilowatt-hours of electricity a year, enough juice to run three million homes and avoid the release of about 23 million tons of carbon dioxide from the atmosphere. This is quite encouraging, but still modest when viewed against the backdrop of the total US energy market: 31 billion kWh represents only 1 percent of the nation's electricity needs.

Historically, investors have been reluctant to go anywhere near wind because it turns the traditional economics of the energy business on its head. Constructing a coal- or oil-fired power plant, for example, consumes 20 percent of an energy producer's lifetime budget; buying fuel and operating the plant gobbles up the remaining 80 percent. Wind and solar power are governed by an opposing rule of thumb: 80 percent goes to costly, up-front infrastructure and 20 percent to upkeep. So yes, the wind is free, but catching it will cost you dearly: A three-blade, industrial-size turbine now costs about $1 million.

At that price, we are no longer talking about wind power as a personal statement by off-the-grid homesteaders. (Residential wind power systems are still quite expensive for the average homeowner: about $40,000 for a 10-kilowatt system.) Instead we're talking about multinational corporations sinking hundreds of millions, even billions, of dollars into wind on the safe bet that fossil fuels will someday run out. They're betting too that the world's governments—not just the folks in Washington, DC—will eventually cough up richer handouts that will make Big Wind an attractive proposition.

Subsidies are a fine tool to stimulate a nascent technology, but the trouble with the boom/bust model is that some players may be tempted to make a short-term investment, then cut bait. Since its inception, Maple Ridge has been passed like a baton to increasingly larger conglomerates. First, it was developed in part by Houston-based Zilkha Renewable Energy, which was purchased in spring 2005 by Goldman, Sachs, the investment banking giant. Goldman formed a new corporation, Horizon Wind, which operates Maple Ridge under a joint partnership with the Portland, Oregon–based PPM Energy (which acquired Atlantic Renewable in 2005) and is owned in turn by Scottish Power of Glasgow.

This spring, both Horizon Wind and Scottish Power were themselves bought by larger international energy providers. Horizon went to Energias de Portugal, the world's fourth-largest wind power producer, for $2.15 billion; Scottish Power was absorbed by the Spanish company Iberdrola, the world's largest renewable energy provider, for $22.5 billion. Lucrative subsidies are drying up in Europe, so it is logical that renewable energy providers would seek holdings in the United States, which is just beginning to get the hang of wind. Noting that the United States is the fastest-growing wind power market in the world, the president of the Portuguese firm told Bloomberg Media, "If you want to be a top player, you need to be in the US."

So far, there are wind energy projects in forty US states, and more coming from the multinationals. For some people, this is cause for rejoicing. Bigger companies can erect bigger projects, pay handsomer royalties to the landowners from whom they lease land, and usher in a world of green power faster than that earnest, mail-order turbine staked out in your organic garden.

But others care not a whit for Big Wind. Watching the flurry of profit-taking, they see not a new kind of energy business but simply business as usual. "This is not about green power," says Gordon Yancey, owner of the Flat Rock Inn on Tug Hill, who opposes the wind project that is literally in his backyard. "There's nothing green about it. The only *gree* in it is greed."

Money makes it all so easy to swallow, and that's the most understandable reason why so many people in Lowville welcomed the turbines. "They've offered us landfills and low-level nuclear waste," says Arleigh Rice, a retired dairy farmer and Lowville's town supervisor, or chief fiscal officer. "I'll take windmills any day." (While those other options promised tax payments and new jobs, they did not offer individual royalty payments.)

During the two phases of the wind farm's construction, each of which lasted six months, every hotel, diner, and restaurant in town was full with four hundred construction workers, a mix of locals and out-of-towners. The wind companies bought up 25,000 truckloads of gravel and 77,000 cubic yards of concrete from local suppliers. The newly built operations and maintenance facility just up the road from the Burkes' house will serve as a headquarters for as many as fifteen to twenty full-time staff employees—a similar mix of locals and out-of-towners—who monitor and repair the turbines. (On any given day, nine or ten of the turbines are down for routine maintenance.)

Rice's counterpart in nearby Martinsburg is an engineer named Terry Thisse. He's a rugged-looking man in a red down vest who sells Jet Skis, ATVs, motorbikes, and snowmobiles from his sporting goods store near the largest local dairy. He explains that in lieu of paying full taxes on the turbines, Maple Ridge's owners will make reduced payments to the county for the next 15 years; according to Thisse, this is a better deal than Maple Ridge would get if it had to pay the presumed property taxes on the turbines.

"This allows us to get a leg up while we're laying out the high start-up costs," says Moore. After fifteen years, Maple Ridge will pay the full tax, and by agreement landowners will never be responsible for any portion. (This is a common arrangement between Big Wind operators and host communities.) The exact size of the reduced payments will be determined by the kilowatt hours produced, but the total infusion to the community is expected to be $8.5 million annually.

The county divides the revenue among towns according to the number of turbines they host. Martinsburg, a town of only 1,249 people, has 102 turbines. Though it typically collects $370,000 a year in property taxes, tiny Martinsburg's first payment from Maple Farm, for the first phase of construction, amounted to $1.13 million. Harrisburg, population 423, with seventy-eight turbines, received $1 million. Lowville, with a population of 4,548, reaped only $110,000, since the windmills there number just fifteen. "Obviously, money like that can do a lot of good in a community like this," Thisse says. "We had to hire a consultant just to help us invest it. The goal is to operate solely on the interest. It's a nice situation to be in. We've got sixty miles of dirt roads in town. So we're probably going to be able to finally pave them." By agreement among all the towns, local schools get the largest share, 40 percent, or at least $3.4 million annually, according to the most current figures. This last disbursement is widely regarded as the smartest investment in Lewis County's future.

A grandmother of two and former middle school principal, Patricia Burke is looking at the big picture. "Our town's population is steady but there's been no growth in years," she says. "We're looking at this as a way to provide jobs and maybe keep some of our children here." Thus far, it has certainly worked for the Burkes: Their son, Robert, landed a job in the operations department at Maple Ridge, and Robert's wife, Jenny, was hired by Moore to run the project's downtown office, just above the JC Penney catalog store.

Still younger residents have also fared well. Not far from the Burkes' home is the sprawling log cabin owned by Kyle and Betsey Burbank. Kyle, a veterinarian, was captivated by the proposal Moore made one day over coffee in the family's kitchen. But Betsey, a home-schooling mom, was fearful that the windmills would be harmful to the couple and their sons. "We went down to visit some of the other windmills. I got some articles and read up on the topic," she remembers. She was satisfied that there was no such danger. "After that, I was gung ho."

Today the family hosts four turbines on their 250-acre farm. On a recent morning, the three older Burbank boys, who range in age from six to eleven, are crowded around the kitchen table, anxious to finish up their math worksheets so they could tend to the Holstein cows they are rearing for the 4-H club. Betsey explains that construction of the windmills was a major event at their home, taped via camcorder and incorporated into her sons' lesson plans on green energy. Perhaps more significant, though, is the fact that the Burbanks are socking away the income from the four turbines—about $26,400 annually—in a college fund for their sons, who now number four. (Newborn Isaiah is upstairs sleeping.) In fact, says Betsey, the boys have each claimed a wind turbine, joshing the others about whose blades are turning—and earning—the most.

Josiah, age eleven, hears his mom talking and leaps up from the table. "Ooh, ooh, can I show them?" he pleads. When she acquiesces, he rushes to the edge of the porch and points to the turbines rising on the horizon above the stark, flat pastureland. The wind is biting. He is dressed only in a plaid shirt, jeans, and socks. "That one's mine," he says proudly, shouting to be heard over the baying of his father's beagles. "That's Abraham's. That's Malachi's. And if you look over here, around the porch, you can see Isaiah's."

For those who oppose large-scale wind projects, Tug Hill has become a symbol of wind gone wrong. A twenty-one-minute amateur film, "Voices of Tug Hill," can be found not only on YouTube but anywhere citizens debate the merits of major wind projects. In the film, a handful of Lewis County residents blame Maple Ridge for a multitude of sins that can never be righted or explained away: the spoiling of their landscape (or "viewshed," in industry-speak), the sound of the whirling turbines, the potential threat to wildlife. Naysayers always say they are not against wind power per se, but against *industrial* wind power, against corporate intrusion into their small-town, pastoral spaces. Anne Britton, an activist and a resident of Brandon, New York, about 106 miles northeast of Lowville, says, "It's about big companies making money. They're laughing at us. They're collecting their money and going back home." Thanks to Britton's efforts, Brandon declared a six-month moratorium on large-scale wind projects last fall and extended it for another twelve months in spring 2007. "We're hoping that if we kick up a fuss, they'll just go away," Britton says.

While Maple Ridge was still in the planning stages, Moore hired Paul Kerlinger, a migration biologist who often consults with wind power companies, to document over four years the number of dead birds or bats found beneath the turbines. Kerlinger says the study is not complete. A preliminary report shows that bird deaths have numbered in the hundreds annually —but that a potentially more worrisome number of bats have been killed. No one knows why. One theory is that bats switch off their sonar while migrating. Maple Ridge is exploring technology that will make the turbines more obvious, or less attractive, to migrating bats.

On the sound issue, Moore says that modern turbines generate less noise—about 50 decibels, the volume of a refrigerator hum— than their clunkier predecessors. But hearing is strongly personal, and what people hear depends very much on wind activity at ground level, where people actually live. Some people swear they can hear turbines on a low-wind day in their homes with the windows open. Some say they hear nothing. Others say they do hear a hum over long distances but have so habituated themselves to the noise that they manage to ignore it.

But Gordon Yancey, whose tavern commands a majestic spot on the plateau, says he despises the whooshing sound of windmills—and this from a man who rents snowmobiles for a living. "Well," he says, "let's look at that. A snowmobile makes a lot of noise but then it goes away. These things are like being next to a freight train that never arrives."

Moore thinks that Yancey and others who object to wind power are in the minority. "This all comes down to an emotional reaction to change," he says. "They don't like the way the turbines look or the way they have transformed their viewshed, so they come up with a semi-scientific argument to sound cogent. The majority of people in Lewis County like the project. But this is all a highly subjective thing." In other words, he could talk himself hoarse and he'd never persuade everyone.

Ultimately, what makes Maple Ridge so interesting and so important is that never before on the East Coast have so many instruments of electrical generation been inserted into the domain of the private citizen. It's one thing to know that your electricity comes from a coal-fired or nuclear power plant six counties over. It's another thing entirely to accept gigantic whirligigs on your own land, to share your life with them, to stake your children's future on them, and to bequeath the financial legacy to your grandchildren.

We make concessions every day for the other fuels and tools in our lives: Oil and coal pollute our water, air, and bodies; nukes leave us with centuries of toxic waste; heck, compact fluorescents contain mercury, even as they lead us to a bigger goal. There's no free lunch here, and Maple Ridge may be an education for us all as we head into the renewable energy future so many Americans say they want.

For Moore, fifty-one and a self-described child of the sixties, this project is part of a larger personal goal that goes back more than twenty-five years, when he wrote his undergraduate thesis on energy regulation. "I've wanted to do this for years, show people that renewables are real, that they are here and now," he says. "Every time I drive to Lowville and see that plateau with the wind coming across it and the turbines catching it, I get a big thrill."

The Production Tax Credit (PTC) for wind power and other green energies expired at the end of 2013.

MARS BOUND

2003

Scientists must overcome significant challenges to take humans to Mars. One of the scientists working on the problem has dreamed of space travel since childhood.

Franklin Chang-Díaz remembers the day his mother told him that a new star soared in the sky. The year was 1957, and the Soviets had just launched a basketball-sized satellite named Sputnik, which triggered the space race. The news captivated the seven-year-old boy, who grew up in Venezuela and Costa Rica. "I was lucky to have a set of parents who paid attention to such events," he says. "That launch was so powerful for all humanity."

From that moment forward, Chang-Díaz lived for space. Cardboard boxes became his rocket ships, his kid cousins his intrepid crew. He listened to news of the first manned spaceflights on the radio. When a museum exhibit touting US efforts in nuclear power came to San José, Costa Rica's capital city, every afternoon Chang-Díaz rushed to the San José International Airport, where the display was located, to learn more about using atoms for energy. In high school, inspired by a NASA brochure titled "So You Want to Be a Rocket Scientist," Chang-Díaz wrote a letter of inquiry, but Houston fired back a crushing reply: NASA careers were open only to US citizens.

"It drove me crazy," Chang-Díaz says. "Even today it does. Why would they encourage us to be rocket scientists if we couldn't be? Space exploration is a worldwide endeavor, and the fact that the United States is on top doesn't mean they should be the only ones in it."

The letter was curt, but it did not dampen his resolve. "It just made it clear to me that I had to come to the United States," he says. "After high school, I got a job in a bank to save money. I told everyone that I was going to go to the United States to become a rocket scientist and an astronaut. Everyone laughed."

More than thirty years later, there is nothing to laugh about. With single-minded determination, Chang-Díaz has nailed every one of his dreams. He has flown seven shuttle missions, a record equaled only by one other NASA astronaut, Jerry Ross. As the first Hispanic American to go into space, Chang-Díaz is regarded as a hero in Costa Rica. He also became a rocket scientist, and his research in fusion fuel physics may well help get us to Mars one day.

At no point in his childhood did any of this seem possible. His mother, a housewife, feared her son would end up in the Vietnam War if he immigrated to the United States. His father, a construction-site foreman and the son of a Chinese immigrant to Central America, had a different perspective: Immigration was a fact of life; you went where the jobs were. Moved by his son's efforts at the bank, the elder Chang-Díaz bought him a one-way ticket to America, and the young man went to live with relatives in Connecticut.

Chang-Díaz pocketed all the money he had saved in nine months—$50—and headed for Hartford, where he quickly convinced school administrators to put him in the high school's senior class. "Of course I didn't speak English," he says, "but I knew it would only be a matter of time. I worked hard at that. I tried not to hang out with Spanish-speaking kids. I got an American girlfriend. As my language skills improved, my grades shot up."

He was awarded a four-year college scholarship, but the day he showed up at the University of Connecticut, he was confronted with a familiar objection. "Everyone was like, 'Oh, sorry. We can't give it to you. You're not an American citizen. We made a mistake.' You see, they thought I was Puerto Rican, not Costa Rican." The embarrassment was enough to persuade state officials to grant him a one-year scholarship.

That was all he needed: He landed a job in the physics lab and worked his way through school. "How's that for luck?" he says. "That was one of the most American things that ever happened to me. Here, there's a mindset that if you work hard, you usually get what you want. It's still a land of opportunity."

In 1977, Chang-Díaz finally became an American citizen. Working then as a fusion physicist at the Draper Laboratory at MIT, he saw this achievement as just another step toward the goal that consumed him still: a career as an astronaut. He had applied for the space shuttle program before he got his citizenship and had been rejected; on this second round, he was summoned to Houston for a battery of interviews and tests. The visit was marked by some fanfare because it was the first time a naturalized citizen had ever been considered for the job of astronaut.

And yet, months went by without an answer. Then one day in 1980, as he sat in a colleague's office describing an idea for a superhot, superfast fusion rocket, Chang-Díaz was paged. "Dr. Chang-Díaz," the caller said, "you have been selected to become a space shuttle astronaut. Do you want the job?"

"What kind of question is that?" Chang-Díaz asks rhetorically, recalling the moment. "I was so excited that I started pacing in circles and wrapped the phone cord around my superior's neck."

Six years later, he flew for the first time on a six-day, ninety-six-orbit mission during which he helped deploy a satellite, conduct experiments in astrophysics, and run an onboard materials processing lab. "I had lots of powerful emotions," he says, recalling that first trip. "When you get to space, as soon as you feel the float, you want to do two things. You want to unstrap because even though you train for zero gravity, you never know exactly what it's like until you're in it. The second thing you want to do is look out the window. That's what blows you away —to see Earth from that point. I've flown many times now, and the feeling is always the same."

Today Chang-Díaz devotes his time to perfecting the engine he first conceived two decades ago. In a lab at Johnson Space Center in Houston, he and his team use hydrogen gas to generate enormous heat. "In space," Chang-Díaz likes to say, "heat means fast."

Zapped by radio waves, hydrogen atoms lose their electrons and are transformed into a plasma as hot as the sun. This gaseous mass—the fourth state of matter found in lightning, nebulas, and stars—reaches 1 million degrees Fahrenheit; it can melt any container devised by man. Only magnetic fields tame plasma. With large electromagnets attached to the exterior of the plasma chamber, his team can lead the hydrogen gas plasma as easily as a pup on a leash, harnessing 1 million degrees of energy to a single unalterable goal: powering a rocket to Mars and beyond.

Chang-Díaz hopes that NASA will use his engine to help boost the International Space Station into orbit. (The station's low orbit takes it through the upper reaches of Earth's atmosphere. The drag slows it down, causing its orbit to decay, so it must be lifted and repositioned.) If the system works well, it will be used for later missions to outer planets. "We measure the performance of a rocket in seconds," he explains. "If the shuttle comes in at 500 seconds, this engine has 30,000 seconds. The space shuttle is unsuitable to go to Mars. It's an oxcart compared to what we can do."

You could say that brilliant, unswerving energy is the story of Chang-Díaz's life. Today in Costa Rica, all schoolchildren know the name Chang-Díaz. His face recently appeared on a stamp, and a pair of biologists named a new species of rainforest beetle in his honor. His mother, Maria Eugenia Díaz de Chang, visits schools to tell how she once lit the flame of fascination that sent a young man to the stars. Chang-Díaz himself works to make sure the young have a chance: He aggressively seeks out graduate students from all over the world to help design the technology his lab uses.

Now, Chang-Díaz may be in his sixties when this planet finally sends a person to Mars, but he still hopes to be tapped for the mission: "Astronauts get better with age."

Chang-Díaz retired from NASA in 2005 to start his own Houston firm, Ad Astra Rocket, which is devoted to developing Chang-Díaz's plasma rocket propulsion technology.

DO MORE WITH LESS

2002

Why do we humans consume so much energy to get around in our vehicles when butterflies get by on a sip of nectar? That question inspired the life of the world's best-known inventor of human-powered aircraft.

Paul MacCready had two older sisters, but they were so much older that he always felt like an only child. Socially awkward, dyslexic, the smallest kid in class, he might have suffered through childhood. But even as a boy, he knew exactly what made him happy. During summers at his parents' cottage, he collected butterflies and moths, marveling at their ability to fly on such delicate wings. There were lots of birds to watch, too, their aerobatics accomplished by such effortless, slight movements. A child with the soul of a poet might have written about that. Another child might have sketched these wonders. MacCready set himself a very different path: How, he began to wonder, could I build a butterfly, or a bird?

Almost seventy years later, it can be said quite simply that MacCready has succeeded: He has built machines that soar through the air with the effortless precision and grace of a butterfly or a bird. There was, just for starters, the Gossamer

Condor, the first successful human-powered aircraft. The Condor was quickly followed by the Gossamer Albatross, which was pedaled through the air and across the English Channel from Folkestone, England, to Cap Gris-Nez, France, in 1979. A few years later MacCready's company built the world's first solar airplane, the Gossamer Penguin, and then a more powerful one, the Solar Challenger, which flew from Paris, reaching an altitude of 11,000 feet, and landed at an air force base in England with energy to spare. In 1987, at sixty-two, he and his team built a bizarre-looking solar-powered car called the Sunraycer, which won a 1,867-mile race across Australia against twenty-two other solar-powered cars. Three years later they introduced the prototype for the battery-powered Saturn Impact, which you can still occasionally find on the road. And when the producers of an IMAX film on the history of flight were casting about for a pterodactyl, MacCready built them an eighteen-foot working replica of the largest flying creature the world has ever seen.

But it is always the human-powered devices we remember, aircraft that seem so delicate, so light, so impossible, yet prove to be as strong and durable as a monarch butterfly migrating from Mexico to Maine. His company built three of them for NASA. All have flown effortlessly, quietly, on solar power, into the stratosphere. The latest, Helios, reached 96,863 feet, two miles higher than any other non-rocket-powered aircraft had flown. It is powered by fourteen electric motors that turn propellers. It will be able to stay aloft for months at a time. The Smithsonian Institution understands this genius. It has collected two planes flown by Charles Lindbergh, three built by the Wright brothers, and five of MacCready's creations.

As a boy he built model aircraft—gliders, ornithopters, helicopters, and autogiros; by the age of sixteen he was through with miniatures, opting instead to fly the real thing. At twenty, bored by powered flight, he chose to master gliders. A glider pilot who wants to stay aloft longer than twenty minutes at a time must search for thermals, as hawks do, exploiting a free ride from nature. By age thirty-one, he had won four soaring contests in the United States and Europe. Now, at seventy-six, MacCready still dreams about carving out time to return to his first love, promising himself: "This year I'm going to get my license and get back into gliding."

Understanding how a light aircraft flies like a bird came in handy when MacCready, the father of three young children and head of a fledgling engineering start-up, cosigned a $100,000 business loan for a friend, who then defaulted. Looking for a way to pay off the note, MacCready began to think about a cash prize offered by British industrialist Henry Kremer to anyone who could fly a human-powered airplane over a one-mile-long, figure-eight circuit. The money—about $90,000—had gone unclaimed for almost two decades. To get it, all MacCready had to do was make a man fly. British engineers had tried in vain, and MacCready's preliminary calculations showed that a conventional design wouldn't work. He worried that figuring out a radical approach would take too much time and be too costly, so he tried to forget about it.

Then, on a cross-country trip with his family in 1976, he and his sons marveled at the flight of large birds overhead, watching hawks and vultures remain aloft for hours without ever flapping their wings. "I think it was the day after the tour of Kitty Hawk when I realized that you could win the prize with a different approach," MacCready says. "The plane would have to be larger and lighter. After that I knew just what to do. There was only one unknown. The wings would be so large they'd need exterior wire bracing, and I didn't know if that wire would have too much drag. So I couldn't wait to get back to the books to figure that out."

Back in California he and a cadre of twenty friends and employees worked weekends perfecting the seventy-pound Gossamer Condor. The design was unprecedented: The plane had a wingspan of ninety-six feet, covered in Mylar sixteen-thousandths of an inch thick. Enclosed in a plastic cockpit, the pilot pedaled to get the bird airborne. The prize was theirs. But no sooner had they won it than Kremer offered another—$200,000 for crossing the English Channel under human power, a distance of twenty-two miles. "He thought it would take another twenty years to do," MacCready says drily. "But we realized we could clean up the Condor's design, up the efficiency by 20 percent, and the pilot could make it across the Channel."

Besides, MacCready needed the money. He'd sunk so much into building the Condor that he'd been able to wipe out only a third of his original debt. So the team headed to England in June 1979 to launch Bryan Allen, a 137-pound cyclist, in a new plane, the Gossamer Albatross. A flotilla of media boats watched as the weather, which had seemed perfect early in the morning, deteriorated quickly. Two hours into the flight, drinking water gone, legs cramping badly, Allen signaled for a tow. But when he maneuvered the craft a few feet higher so that rescuers could hook a line to the Albatross, he found calmer air and waved them off. He pedaled another forty-nine minutes—sometimes with one leg, sometimes only inches above the waves—and touched down in France. "He did a truly phenomenal job," says MacCready, who was nursing a broken foot and watching from the deck of the lead boat.

By now MacCready was in the grip of an obsession—squeezing every bit of efficiency from aircraft, pushing engineering to new limits. He rarely paused in his work long enough to consider why he cared so much about such things. But then, in 1982, he had to make a speech in acceptance of the Lindbergh Award. In his research for the talk, he stumbled upon a story that spoke to his heart. "Few people know that in the last twenty years of his life, Lindbergh dedicated himself to the environment," MacCready says. "In Kenya, Africa, he found himself looking at a hawk, and he had to ask himself which would make him sadder—if there were no birds or no airplanes? And he chose birds, even though he was an aviator. And that helped him resolve to work on environmental things."

MacCready reflected that his own work was an expression of his lifelong fascination with wild creatures. The realization helped him clarify the goals of his company, AeroVironment, which now focuses on developing alternative energy sources. They manufacture batteries and chargers for short-range gasoline-free vehicles such as motorized airport carts and forklifts.

They crank out half-ounce robotic planes that are used for reconnaissance by the Department of Defense. And they are perfecting solar power and hydrogen fuel-cell technology to keep NASA's Helios aloft indefinitely, carrying atmospheric monitoring equipment. The team also finds time to tinker with one of MacCready's pet projects: a human exoskeleton whose robotic legs would give wearers the lifting and climbing power of a Titan.

All of these endeavors dovetail neatly with MacCready's credo: Do more with less. What worries him these days is that the rest of us seem hell-bent on doing more with more. "I'm basically pessimistic about the future," he says. "There were 1.7 billion people on the earth when I was born, and now there's 6.2 billion. Almost four times as many. If you look at all the vertebrate creatures on sea, land, and air, humans and their livestock and pets occupy 80 percent of the mass of vertebrates. That's an overwhelming percentage. Natural animals account for the rest. I'm very dismayed that we are heading into a period where we may not have wild animals around. I show curves when I give talks to get people thinking about this: Humans, livestock, and pets keep going up, and all wild nature is going down."

For MacCready, doing more with less means weaning ourselves from fossil fuels. It means investing in technologies that harvest sunbeams and wind and charging the batteries in home appliances with power drawn from these sources. It means building cars that guzzle hydrogen and expel harmless water vapor. And it means pursuing those technologies today—so that we're not rudely awakened when nature pulls the plug on fossil fuels. He is appalled that such planet-altering issues are a low priority on the national agenda. "It's got to get so bad that the politicians respond," he says. "It's a devastating problem for the United States to have key officials—the president, the vice president—associated with the oil industry. There are lots of people working on good, all-natural energy sources. Unfortunately, the politicians believe what only 1 percent of the scientists believe."

If MacCready sounds exasperated, it's because he cannot bear to lose what inspired him as a child. He has always known that his machines are, at best, facsimiles of organisms that nature got right the first time around. Every bird, bee, and butterfly in the kingdom of life flits at the edge of what is aerodynamically possible, but still succeeds. MacCready has lived a life playing on that edge, and he wishes the rest of us would join him there.

MacCready died in 2007 at the age of eighty-one. His firm, AeroVironment, continues to follow in the footsteps of its founder.

KNOW THYSELF

2000

The assignment for a science magazine's twentieth birthday in 2000 was simple. Talk to a bunch of scientists and ask them what we need to know in 2020 that we don't know now. Some of their predictions already seem quaint.

You know things, you child of the twenty-first century. You may not stop to think about it, but you know stuff to get along in the year 2000 that your hallowed progenitors could never have dreamed of. You know how to delete. You can pull down a menu. You know how to change the channel on your TV without getting out of your overstuffed chair, a luxury your grandparents did not enjoy. You know—or should know—how to keep your kid from downloading porn on the Internet. You even know the word download. You know other words too. Scary words. Ebola, mad cow, West Nile virus. At the very mention of these words, your mind knows to give you the creeps.

As always, knowledge is power. What you know gets you through your day. Protects your family. Keeps them safe. And what you need to know has changed—a lot. Twenty years ago, you took notes with a pen, not a pointer pushed across the face of a personal digital assistant. Twenty years ago you still thought a mouse was just a rodent. Had some prescient soul sidled up to you on the street in 1980 and said, "Listen, buddy, soon you're going to need to know how to operate a big glowing box on your desk by sliding a plastic thing around," you would have seen him as a madman, not a prophet.

By the year 2020, you will need to know stuff you can hardly guess today. You've heard all this before. You're hip to this technology thing. You know the drill: Gadgets change; you adapt to the gadgets. You learn new buzzwords, talk the talk, and keep on going as humans have for millennia. But in doing so, you reduce technology to a heap of glorified hand tools, the equivalent of sticks chimps use to extract ants from a hole.

In twenty years it should be painfully clear that technology never just hands us tools; it grants us a passport to a world where choices multiply, desires are ignited, and new moral decisions confront us. Like an odyssey-starved traveler, you'll wander through a kind of exotic street fair, senses assaulted by too much information. You'll find yourself frazzled just trying to draw a bead on all the options, while a numbing calliope tune insists it's all great fun. A lot of what you'll encounter will just be cool, requiring nothing more than nimble minds and fingers. By the year 2020, for example, you will need to know how to talk to your house. Today your home contains dozens of appliances, each working independently. But someday you'll cross the threshold and everything will know you're home. The lights will flicker on, the air conditioner will have kicked in, the refrigerator will clamor to enumerate all the meals you can assemble with the groceries cached inside. In exchange for this convenience, you'll share your abode with a horde of circuit-heavy infants that constantly burble to each other and cry out for your care. If you're the one with mechanical flair at your house, when you come through the door at night, your spouse will be pulling you aside to whisper, "Honey, I need you to talk to the robot."

By the year 2020, you will have to learn to drive a more automated car. You'll get behind the wheel of a smart car that avoids fender benders by braking before you even see danger looming. At a much later date, you will slip into a bucket seat as if at the movies—snacks, reading material, and sodas at the ready—sit back, relax, program the car, and over the freeways to Grandmother's house you'll go. Like the toaster and coffeemaker back home, the car's sensors will monitor the activity and destinations of other cars on the road. "Going my way?" your vehicle will bleep in autospeak. "Indeed," responds the living room on wheels in the left lane. And the two will hitch up and rocket toward their common goal together. This technology will conserve fuel and may save lives, but the pleasure of driving as you know it will be gone. That's something you should know.

But perhaps by now you've realized that for every convenience technology bestows upon us, it chips away at something else. All of us, great souls as well as lost ones, must in time wrestle with this notion. If you are the poet Blake, in metered rhyme you decry the Satanic mills; if you are Kaczynski, you take to the hills and spit death by snail mail. Most of us simply acknowledge the trade-offs and move on.

Each time we do this, though, we march farther away from a world we can touch and comprehend in our bones toward one that we pray will work better. Consider: In the year 2020, you'll identify yourself, gain access to homes and businesses, and board aircraft after a laser has measured the shape of your irises. But the price will be loss of privacy. A record of your transactions, your daily comings and goings, will be just a keyboard tap away from others.

Booting up your home PC has already become a public act. Meander the web today, and almost every move you make is cataloged in service to the gods of commerce. They know what you're buying. What you listen to. Where you chat. By 2020 you'll need to know how to clean up that electronic trail day in and day out. "Say you were searching for information on hats," theorizes Jaron Lanier, computer scientist, musical composer, and virtual reality pioneer, "and you saw a link about hats, but when you got to it, it was actually a weird pornography site about hat fetishes. Then it turns out there's a record that you visited this site, and now you're getting bombarded with offers from people with hat fetishes. Furthermore, your friends are being contacted in case they have hat fetishes. All of a sudden you're the hat fetish person in your social circle, and you have to go in and undo it."

To throw the hounds off your scent, Lanier says, you could spend the afternoon downloading the Great Books or posing as a do-gooder in search of charities deserving of your drachmas. In time, you'll be wielding electronica for the same reasons medieval crusaders took up sword and lance: to ward off intruders. Rooting out destructive viruses and spam in your equipment will become old hat, as will the regular checks you'll be performing on your groceries and yourself. Tomorrow's Kaczynskis will be able to concoct harmful viruses and insinuate them into the food supply, or perhaps release pathogens in public places. You'll need to be ready for them. Daily computer checkups of your blood, saliva, or bodily waste will be effortless, the medical equivalent of checking your stock portfolio. "Real-time monitoring," says James Weiland, assistant professor of ophthalmology at Johns Hopkins, "will tell you in the morning what vitamin your body is low on and what to have for breakfast."

With all this new information, you'll stand a better chance of living well beyond your biblical allotment of threescore and ten. More than 200,000 centenarians will inhabit the United States in 2020—why shouldn't you be one of them? To reach that age you'll need to know enough to make more complicated medical choices: Do I want to jettison a limb and wait five years to regrow another? Shall I allow a phalanx of nanobots to scrape the plaque out of my arteries or opt to replace the vessels altogether? "Amateurs may be fooling around with black-market genetic manipulation," says Marvin Minsky, one of the founders of the Artificial Intelligence Lab at MIT, "maybe extending their lives by lengthening their own telomeres, the ends of chromosomes believed to control life span. Or they might, in fact, be growing new features in their brain."

By the year 2020, science will understand the Creator's software well enough to tell you a great deal about the genetic hand dealt you and those you love. Science may even help you decide if you should quit loving them. These days it's not unheard of for one partner to investigate the other's background or assets before marrying. In the future you'll need to access your betrothed's genetic map, see what diseases he or she is likely to contract, assess the appearance and health of your children, and perhaps even size up your love's mental health. Of course, this swings both ways. In this world, you will be forced to ask: Do I want to know if I'm earmarked for heart disease or breast cancer? Do I want my potential spouse to know? If I know this, and my doctor knows, does it mean that my insurance carrier must know? If this last one scares you, it should. It could mean the end of health care as you know it.

This is just the beginning. Once we know the future, we're going to be tempted to rewrite the software. Clearly, it would be an act of kindness to reach into that fragile, permeable, four- or eight-cell being and rid it of the disease that cut short the life of its great-grandfather. But why wait for conception? Why not design your kid, toes up, out of whole cloth, using the blue-eye gene, the blond-hair gene, and the excel-at-lacrosse gene? Ban such tinkering, and citizens will merely scurry underground in order to conceive the perfect child.

If we can tear ourselves away from such selfish goals long enough to look around, we will have to face the fact that technology favors some and eclipses others. Bill Robinson, who spent thirty years as an electrical engineer with Canada's Nortel Networks, has been thinking about this issue recently. "We spend our time and effort creating exciting new communications technologies," he writes, "yet half the world does not have access to a telephone. We use the Internet to order the latest novel, yet many people in the world don't have access to books. We are now discussing embedded processors to connect our refrigerators to bathroom scales and the grocery store, yet many children in the world go to bed hungry at night."

This grisly reality will be harder to hide from when our planet swells to eight billion people in 2020. For Lanier, the most heartbreaking scenario is festering in the third world, where, he believes, the current generation of children—lacking food, lacking skills, lacking aid, lacking education—will be lost in the next techno-revolution. "What is going to happen to all these people as they start to age, say, twenty years from now?" he wonders. "You're going to have to somehow live while you watch a billion people starve, which is going to be a new human experience. How will we do that?"

Good question. And just one of many difficult questions waiting. How can I choose between two genetic scripts for a child I have yet to know? How much of myself should I reveal on the web? How will I cope with all these machines when they break down, including the self-replicating nanopests that may be residing in my flesh? In our zeal to be happy little technologists, we'll turn, much as we do today, to the web for answers. And we'll perfect the art of being disappointed.

If any medium ever resembled the human unconscious, the web is it: a place of hidden wonders, stray inane thoughts, peaks of brilliance, valleys of perversity. And no apparent governor. Type your query, hit return, and voilà!—10,000,000 hits. Good luck shaking them down. Even in 2020 you will always need to know if the facts you've dredged up are accurate and truthful. With so many sources doling out information, you will need to know: What is this person selling, and why is he selling it? Most unsettling is the fact that these precious touchstones are not permanent. They never will find their way to the library stacks. Instead we are moving closer to Orwell's nightmare: the truth ceaselessly modified, altered, edited, or altogether obliterated. Here today, gone tomorrow, with nothing but a bewildering ERROR 404 FILE NOT FOUND left in its place.

By then, you will no longer be a child of the twenty-first century. If anything, you'll be an elder, your mind and body augmented, your chromosomes refreshed, flexible computers woven into the four corners of your garments. On the one hand, your workload will multiply as you bat away each glitch resulting from the increased number of gadgets in your life. On the other, you will be forced to take on moral questions no human has ever faced. When will you find time to do that? How will you contemplate when everything is speeding up and time for reflection is practically nonexistent?

That's you in twenty years. Like the machine that inspired your age, you will be constantly scanning, processing, sifting, searching for a code to guide you through. And yet the key, the compass, the answer, was once offered in a temple at Delphi. What will you need to know in 2020? *Yourself.*

NOTE: AMAZINGLY, NONE OF THE EXPERTS WE interviewed predicted the events of 9/11, though many worked on national security projects at US national labs.

WORSHIPPING AT THE CHURCH OF THE OPAQUE

2002

This story began with a simple lament, uttered by an engineer: "We're going to be in trouble in a few years because kids don't take things apart anymore."

1. On the Shoulders of Giants

Three decades before he invented one of the technologies that made the Internet possible, Leonard Kleinrock begged his mother to take him to buy parts for his first homemade radio. The year was 1940 and Len was a curly haired kid of six. In a Superman comic book, he'd found plans for building a crystal radio. In no time he rounded up what he needed: a used razor blade from his dad's shaving kit, a cardboard toilet paper roll, the tip of a lead pencil, some copper wire, and a telephone headset filched from a public phone booth. There was just one more item, not easily found at home. To buy it, his exasperated mother hauled him a hundred city blocks by subway, from their neighborhood near Harlem to Canal Street, New York's mishmash of junk stores, rag shops, and spare parts emporia. Into one of these places Len walked, his shoulders squared like a prizefighter's.

"Whaddaya need, kid?"

"I want a variable capacitor!"

"What size?"

The guy had him. Len had no idea what size. Hell, he had no idea what a capacitor was. "But they helped me," remembers Kleinrock, sixty-seven, a UCLA computer science prof who in 1969 transformed his lab computer into the first node of the ARPANET, the precursor to the Internet, and sent the first web message ("L-O," or "login") to a colleague at Stanford. "I told them what I wanted, they sold me the part for a couple of cents, I went home, put the damned thing together, and it worked. I had music. I was hooked!"

A few years later, a kid from the Bronx named Malvin Carl Teich kicked up a fuss in his parents' kitchen. He refused to visit his grandmother without his favorite toy: his mom's electric broiler. "I made my mother put it in a box," says Teich, sixty-two, a photonics expert at Boston University. "I wasn't going to let her go unless she let me take this damned thing with me on the trolley." At Grandma's house, six-year-old Mal spent a happy couple of hours dismantling and reassembling the broiler. "I didn't do anything bad to it," he says. "My mother could still use it when we got home."

Around the same time, in the Pittsburgh suburbs, a kid named Robert Lucky was playing in the woods when the trees thinned and the boy found himself standing at the edge of a towering, sodden pile of wires, brass bells, and heaps of black plastic. "It was a telephone boneyard," says Lucky, sixty-six, who today works in the telecommunications industry, renowned among electrical engineers for his insightful lectures and articles on technology. "It was a dump. There were old switches and parts and who knows what kind of equipment. I spent days there, taking stuff, and it was just really great."

Once there was a time when kids took things apart. They lay waste to home appliances and vehicles, scavenged parts in back alleys, ordered mail-order electronic kits, and used what they found to build cool stuff like motorized go-carts, crystal radios, and sound speakers. Every vacuum tube, transistor, or gear they touched with their hands flipped a switch in their hearts and minds that still burns today. Besides learning how machines worked, they gained confidence, coordination, and an appreciation for the beauty of a well-made machine. Many of those kids became scientists. Thirty, forty, fifty years later, they look upon those days with bittersweet longing, because technology has changed, and kids don't take things apart the way they used to.

Some scientists say it doesn't matter. They think that kids today are just as talented and creative as they were in the days of the $8.75 radio kit. Others say it's a terrible thing to miss out on doing science with your hands, and even worse, they warn that something essential will be lost to our culture if kids don't get this early training. Mostly, it is engineers who say this, in parts-cluttered labs where they're building all-optical routers, or between classes at the university after some sophomore admits he can't solder and a piece of the prof's heart splinters. Now they're all scrambling to impart this knowledge to young people before it's too late.

Integral to the engineers' lament is the notion that there is something irreplaceable learned when hands meet a three-dimensional object. Frank Vacanti thinks he knows what that is. He nurtured his mechanical competence in his bedroom growing up in Nebraska, surrounded by dozens of spare parts. He scared his parents to death by endlessly removing and replacing electrical plugs on their appliances. Later, less dangerously, he built his own stereo equipment and circuit boards. He thought he'd become an engineer but instead became a research physician who dreams of ways to end strokes, halt vascular clotting, and repair spinal cord injuries. He maintains those early lessons serve him well today. "When you took something apart you saw that it was not as complicated as you thought," he says. "Each part does a job and you could figure out what that job was. When you're doing research, you want to predict an outcome, and think ahead to prevent failure. When you've had a lot of experience tinkering, you get used to failure. Tinkering and failing now helps you later. If you don't do it early in life, you're gonna fail later. Boy, if you can discover how something works, then you carry that experience away with you. You have a sense of accomplishment."

Satisfaction came easily once because machines were easier to fix. You could open them and visually inspect their innards. When the insides were exposed, you could diagnosis the problem, ask your buddy for advice, do a little trial and error, and get it working again. When the young Teich watched his father, an early television repairman, remove the Masonite board behind a TV set in the mid-forties, father and son looked upon a dusty jumble of vacuum tubes—glass tubes which were at that time the only way to transmit electrons. If a tube went bad, you replaced it as easily as popping in a light bulb.

In those days, if you wanted a stereo, you built it. You bought a high-fidelity push-pull amplifier from Heathkit, a company in Benton Falls, Michigan, for $14.95 and put it together nights and weekends. "The reproduction quality is truly superb," gushed the company's newsprint flyer from the fifties. "You hear the breath of the singer and the colorful harmonies of the musical instruments." This was consumer electronics at its most egalitarian: rich or poor, teenager or adult, you could not do much better. "I'll never forget the thrill when you first opened your Heathkit box and saw all those parts," recalls Lucky. "There was something tangible and sensuous even about it. Even the smell from the box and those parts."

Later, if you were getting serious and had the money, you could ascend to the finest electronics in the land—government surplus. You prowled the cluttered haunts of scrap dealers who sold what IBM or North American Rockwell didn't need anymore. When a piece of technology became dated, Uncle Sam or the companies in his employ dumped it, and it was sold like used furniture. You could pick up old airplane parts, optical systems, parts of missiles, rockets and the computers that ran them—hunks of electronic equipment once worth hundreds of thousands of dollars—for $100. Peel back the skin and you could investigate an exquisite architecture. You learned how the masters laid out circuits, propagated signals from them, and sheathed it all in an artistic wrapper. At heart this was a language, and you learned to become fluent in harmonious design and function.

This is hard to appreciate in an age when so many household appliances look disposable from the moment they leave the factory. But once upon a time, you paid a handsome price for a handsome machine and expected to keep it for a long time. As a result, the radio in Grandma's living room, with its Gothic windows and burnished cherry wood, was a synthesis of art and science. She saved her money, chose it carefully, and brought it into her home and life. When it broke, she had it fixed. It was an investment. In the world the pre-transistor engineers remember, form was always indivisible from function. From T-birds to P-51 Mustang bombers, the workmanship outside said as much as the workmanship inside. Rooting around in the guts of these machines, you gained an appreciation for both.

After 1948, technology moved a step away from glass tubes but the work could still be eyeballed. Thanks to a trio of guys at Bell Labs in New Jersey, tubes were being replaced by quarter-sized hunks of germanium studded with gold wires. Transistors were elegant as hell, the most efficient switch man had ever devised. All day long they'd spew electrons across an intersection making a single decision: Positive, negative. Yes, no. On, off. One or zero. Transistors didn't heat up, didn't buzz, didn't shatter if dropped. They did their job and shut up. Electronics was a delightful puzzle: You soldered ten or fifteen coin-sized components such as transistors, resistors, diodes, capacitors—all various conveyors of energy—to pegboards to build a logic circuit that did a specific task: turn a radio on and off, heat the waffle iron, or amplify sound.

The world of hand and eye began fading after 1959, when scientists began their mad dash for power and speed. They ditched pegboards for silicon, and packed dozens of increasingly smaller logic circuits onto a single wafer. More transistors equaled more power, more tasks accomplished. Now technology is entering the realm of nanomachines, where the scale is a billionth of a meter. An adult's thumbnail can cover an integrated circuit containing millions of transistors. The wires connecting transistors are finer than human hair, fast approaching the size of electrons themselves. Transistors still do the same job—on, off, on, off—you just can't see them do it. The guys who design the chips don't touch soldering irons. They type. If they want to get fancy, they use software that generates an image not of the actual chip, but the theoretical one, the one they are dreaming of. The real one doesn't exist until they hit PRINT.

The faster computers became, the faster those great old learning tools dried up. The lifespan of a government computer shriveled from twenty years to two, so much of the surplus market disappeared. Heathkit kits vanished. Who builds stereo components when you can get a CD player for under $100? Kids opened the backs of radios and toys and yawned. There was nothing to see. Nothing to fix. Nothing you could stick a screw-driver into. Nothing to solder. So they didn't. They weren't fools. With computers, the genius wasn't in the hardware anymore; it was in the software. Power lay not in the tool but in how you used it. One by one, kids grafted themselves to the machine. Who could blame them? All of us did the same.

What a world this is, not of hand and eye but of mind and machine. Imagine if someone had swiped Michelangelo's brushes in the Renaissance and assured him he could paint better simply by thinking—and he proceeded to think the Sistine Chapel into being. Teich and Kleinrock and Lucky did just that. They grew up playing in a world before transistors and later coaxed a new world into being, but gave up touch in order to do it. "I really lament it," says Lucky. "I enjoyed doing that stuff. I miss it. There was a connection to the physical world that was almost essential back in those days. The good thing about it was that it pulled me into engineering as it did my entire generation. Those were the things that attracted us."

But now, as a grown man, he feels it is entirely appropriate to inhabit the virtual world. "In some sense," he say, "it's a handicap to work too close to the physical world because it pulls you toward a simplicity and you've got to work at a higher level if you live in a complicated society. You can't be worried about that distributor cap anymore. I asked my wife once, 'Doesn't it bother you that you don't know how a television set works?' And she said, 'Oh, but I do. You turn the knob here and it comes on.' Unfortunately that's the way we have to look at the world today. There's too much complexity behind the knob, and you have to worry about the knobs and not the individual transistors in back of it."

American schools have rushed to give kids the keys to that virtual society. The last two decades have seen an unprecedented emphasis on math and science, sparked largely by a scathing 1983 government report, "A Nation at Risk," that found American school kids lagging behind kids in other industrialized nations.

A committee, which included Glenn Seaborg, the Nobel physics laureate, recommended renewed emphasis on math, science, and computer skills. Hands-on learning—a teaching method first advanced in the Sputnik days—made a comeback. Hardly earthshaking, hands-on math and science lessons exploited the fact that children under twelve are concrete learners, most readily absorbing concepts they investigate for themselves with real objects at hand. But this new initiative could not counteract a larger phenomenon taking place in American society. TV and computer games were rapidly replacing outdoor activities, sports, and other hobbies that occupied kids in earlier eras. To make matters worse, schools were acquiring computers and modems, sometimes sacrificing library and science lab budgets to do so. "We should have been adding computers to the repertoire," says Harold Pratt, onetime president of the National Science Teachers Association, "not replacing subjects with computers."

Today Pratt and his colleagues fear that we have reared a nation of passive learners, and college professors say those fears are not unfounded. "They're good with a keyboard and mouse. They think everything is on the web," says Ira M. Cohen, a mechanical engineering prof at the University of Pennsylania, of students he encounters. "They'll make a mistake in calculating something quite physical to 10^{12}—something the size of the universe as an error—and they don't recognize that it's an error. 'That's what the calculator said,' they say."

Stories like this abound, and not only in technology circles. "The information has to appear in a video or on a computer screen before they really seriously believe it," says Jeff Schaeffer, a wildlife biologist who teaches environmental science at Eastern Michigan University. "The data have to be processed for them and given to them."

In the mid-eighties, enrollment at engineering colleges did plummet, but the kids starting coming back in droves a few years later. Richard Heckel, an engineering prof at Michigan tech who for twenty years has crunched stats for academia and industry, attributes the cycle on market forces. When jobs are tight in engineering, kids aim for the "soft" sciences: biology, sociology, psychology, and economics. (Vacanti, the anesthesiologist, made roughly the same decision in the seventies.) In three years, predicts Heckel, who is sixty-eight, engineering schools will be screaming again for relief. Technology jobs are among the hottest around, a fact which has not escaped high school seniors and their parents. Reared on two decades of computers, Heckel adds, young people are more than ready and eager to devote their minds to the glory of the Beige Box. "I don't buy the idea that kids are not able to do what we used to do," says Heckel. "Success doesn't have much to do with soldering Heathkits. Kids are coming today with computer experience. They want to build stuff, make stuff, or rewrite code. They're very inventive. I have great faith in the youth of America. Maybe I'm naïve, but I think they're great."

Other profs beg to differ. "It's hard to find good hardware hackers these days, even at MIT," says Joseph Paradiso, a researcher at MIT's Media Lab who built music synthesizers as a kid and is today designing the first generation of wearable computers. "We have to teach them those skills more often than we did before. Some still learn this way, others take out these instincts hacking software. It's still in the culture. I've shared these sentiments with others and most agree. Certainly hardware hackers are going for a premium these days." Confides Paradiso's

MIT colleague, Marvin Minsky, regarded as the father of artificial intelligence, "I once found a student who didn't know which way to turn a screw. I think it's a very serious problem. People without mechanical skills are handicapped."

In a sense, Minsky's brush taints us all. American life, as it is lived today, does not require many of us to cultivate competence in mechanics—or husbandry or cooking or sewing—the way it did, say, in 1950 when 40 percent of us still lived in rural areas. Today, the skills not mastered in childhood are picked up later, on the job, at school, in a battery of night school classes. But if one aspires to be the next Dean Kamen, to give the world a zippy scooter—the Segway—it didn't know it needed, they've gotta feel this stuff, as Kleinrock says, in their *kishkes*. Their guts.

Bran Ferren, a computer scientist, designer, and inventor who consults with technology firms, has bandied this idea around with his elder, Minsky. He too has noticed a deplorable lack of training in a specific area. "There's a generation of engineers that is suffering from not having grown up with surplus, because their brains did not form with the physical interaction of advanced technology from an early age," insists Ferren. "I think this is a real phenomenon...I think we're also just beginning to see this. The generations of people that grew up with this stuff are just starting to retire now, so as a result you won't see the result for a few generations. But I can tell you if you talk to people—whether it's people with students or people attempting to hire workers—there's a fundamental issue of finding people who have the skills to do the design you want."

Yes, you can teach someone all these skills, but it's like teaching someone to paint who has never touched a canvas, to write who has never scribbled a line. "If people have not gotten into this prior to getting into college, they'll never really get that kind of background," says Jeff Harrow, fifty-two, a nationally-known technology consultant and author who grew up building his own TV and operating a ham radio. "I'm sure that the current engineering courses teach students the rudiments of that, but I don't believe they develop a passion for it if they're just taking it for a course."

And where does passion begin? The eyes and ears alone are passive senses incapable of influencing the objects we encounter in our daily lives. But the hands grapple, manipulate, revolve, and pull the world in for our inspection. How can you build electrical circuits without ever accidentally experiencing a jolt of electricity? How can you harness the ether when you've never felt its buzz? In his 1998 book, *The Hand: How Its Use Shapes the Brain, Language, and Human Culture,* the neurologist Frank R. Wilson observed, "When desire prompts anyone to learn to do something well with his hands, an extremely complicated process is initiated that endows the work with a powerful emotional charge."

Forty years ago that passion was hard to miss. It did not seem unusual then for an engineer to be concerned with the whole package—how a machine looked, felt, and operated. The earliest designers of the Corvette—Harley Earl, Zora Arkus-Duntov, Larry Shinoda—were both stylists and mechanics. They admired the lines and engines of European roadsters and did them one better. But how do you learn about beauty when you're grafted to a Beige Box? Well, clearly you don't. The fact that we moderns had to rediscover the word ergonomics should embarrass us. "I think we're losing an aesthetic in the design of machinery," says Ferren, a burly, bearish engineer who in his spare time is trying to build a car that will tackle any terrain: up mountains, across deserts, through snow and ice. "If you look at the typical personal computer it's like the design project from the institute for the aesthetically challenged. They don't work well. Do you think a hundred years from now the PC you use now is going to be an art object coveted by collectors? But if you buy a wood plane made a hundred years ago, it is. If you buy a drill, if you buy a sewing machine, if you buy the best technology developed a hundred years ago and looked at it, they're joyful objects. They were physical things which were the combination of human creativity and intellect, but they transcended merely form follows function. They were devices that were beautiful. How many of those things are produced today?"

To go from mind to machine, skipping hands, is not only

joyless, it's dangerous. The best theoretical scientists who influenced the course of the twentieth century were like da Vinci in his day. George Gamow, the father of the Big Bang theory, took frequent breaks from physics to mess around with children's linking blocks. To passersby, his projects resembled meaningless towers, but to Gamow, whose imagination was boundless, they embodied elaborate protein chains. He corresponded frequently with Watson and Crick, and is today recognized as one of the unseen contributors behind the pair's discovery of the DNA molecule.

Claude Shannon, who as a grad student recognized that all information—telephone gossip to Beethoven's Ninth—could be reduced to a litany of ones and zeroes, always urged his underlings to practice illustrating their mathematical theories with a "homely physical example." Shannon, the father of binary theory, was never seen without a Swiss Army knife in his hands. Each night he patrolled the halls at Bell Labs on a unicycle of his construction, designed to let him juggle and ride at the same time.

Albert Abraham Michelson, the first American to win the Nobel Prize for Physics, clocked the speed of light with an interferometer he built himself, and still excelled at music, painting, and billiards. Einstein hailed him as the "artist in science."

2. The Death of Transparency

So today, when engineers pine for the old days, they worry not that we won't have scientists or engineers, but that we'll just have average ones. They feel like they've got their work cut out for them: They've got to figure out a way to teach the virtual generation how to do things in the real world. They're meeting too many college students who say they want to be engineers but who have never taken apart an automobile engine. Students who know how to hack software but don't know that you shouldn't touch the tip of a soldering gun. Students who possess an uncanny facility with calculators and spreadsheets but who can't do mental math well enough to scribble rough estimates on the back of an envelope.

And their parents don't seem to get it either. In fact, they are usually pissed when they call David Kelley. "Why," they demand of the engineering prof, "why does my kid need to learn how to weld? He's not going to some *trade* school. This is Stanford, for Pete's sake!" If they're not sputtering too much, fifty-one-year-old Kelley will try to get a word in edgewise. Engineering is not just about thinking and playing with computers, he'll patiently explain. It's about thinking and doing. Tearing stuff apart and putting it together again is how good engineers get to be great—great, that is, like Kelley, who gave the world the first Apple computer mouse, the Palm V Pilot, and a raft of other wonderful devices.

Now, granted, every generation thinks it alone held the keys to the kingdom. So some of this griping is classically human.

Still...

"They've totally lost the sense of the size of a number," says UPenn's Cohen. "When engineering students had to use slide rules, they had to understand the number because the decimal point was in your head. You supplied the decimal. When we went to calculators, we got results with precision, but the magnitude of the number is off by a horrendous amount. Last semester I taught a class—Intro to Flight. Kids needed to calculate the wingspan of an airplane. They got results like 5,000 feet. I'd ask them, 'Ever see an airplane with a wingspan of a mile?' 'But that's what the calculator said.' 'Did it occur to you that when I wrote that problem I was not insane?' This happens in every class. If it were money, I think they'd understand the difference between 10^{-6} and 10^{12}. Which would you rather have: 10^{-6} dollars or 10^{12} dollars? I betcha they'd know."

You might think it's foolish to carp over a few math mistakes. But in engineering, everything outs. In a sense, a misplaced decimal point is far more dangerous than a misplaced comma.

Woodie Flowers has been railing about this for thirty years. Rail-thin, with a mustache, glasses, ponytail, and a neck like a peacock's, Flowers is MIT's answer to Rube Goldberg. He started welding at age ten in his father's repair shop in rural Louisiana, and built the first car he ever drove—a hot rod roadster—because, as he says, "What else is there?"

Flowers has been MIT's staunchest advocate for real-world, nut-and-bolts, nicks-to-the-hand training for all engineers. For a while, Flower taught a course called "How Things Work." He'd walk into a class, hold up an automobile valve and ask, "Anybody know what this is?" A heated guessing game would ensue. He'd demo common household items, office machines, appliances. He'd peel back the skin on a vending machine, say, and show them what happens when the coin drops in. He'd take apart a clock, teach them how gears work. His rapt audience grew from a dozen kids to more than eighty before the class had to be split in two.

In that class and others, he would be reminded again and again how much there was still to teach. In essence, the mysteries of machines constitute a language, and these kids didn't know how to speak it. They just didn't have the words. In the shop they would pick up a piece of aluminum and call it steel, or draw no distinction between PVC and Kevlar—to them it was all plastic. If Flowers said, "Don't worry, we'll use a ratchet," the inevitable response would be, "What's a ratchet?"

Flowers loves computers as much any pimply faced frosh. He loves pop culture. Digs those little *Star Wars* robots. Understands the allure of video games. And knows that hacking into your dad's online bank account gives you a bigger rush than building a crystal radio. But he also knows that computers and software tend to isolate the user in a rather abstract realm. "Good engineers," he says, "are able to say, 'Whoa! That doesn't sound right! That's crazy. There's something wrong here.' And they try to find out what's wrong." Having that skill is especially critical these days, because a design can go from great to incompetent in a matter of clicks. If you don't know what you're doing, Amtrak trains don't run, buildings collapse, oil rigs pancake into the Norwegian Sea, and hundred-million-dollar orbiters smack into Mars—twice.

"Unless you have touched things, unless you have taken

things apart, unless you've had experience using tools, you're probably at a disadvantage," Flowers says. "The virtual world has taken the place of much of the physicality, and students don't recognize that, say, instantaneous acceleration is not possible in the real world. One of the things I say to help describe the issue is that Mother Nature applies all of her rules all the time; there's never a moment's lapse in her diligence. And people who are physically naïve make naïve designs. If you can't map that into the real world, you're potentially dangerous."

Once upon a time, before David Kelley ever knew he was destined to design airplanes, cash machines, the Polaroid I-Zone camera, the world's first stand-up toothpaste dispenser, or even that Apple mouse, he was just a little kid growing up in 1950s Ohio. In his memory, America was a collection of blue-collar neighborhoods where everyone messed around with their cars or built their own stereos from Heathkits. When Kelley was bored, which was often, he took stuff apart: "I remember bolting two bikes together, taking the front wheel off one bike and putting it on the back of another bike, and seeing if two people could ride them." (They could.) "That's what you did. It's the same as what you did to make money. You had a paper route or mowed lawns. It was all physical somehow. Maybe you watched a little TV depending on how strict your parents were. You were out of the house doing something. There was no sitting in the house playing computer games or even studying."

You don't need a historian to tell you that times have changed. America's growing affluence played a role: Why fix your own car when you can pay someone else to do it for you? Why build a stereo when you can buy cheap components? But that doesn't explain it all. The machines themselves have changed, and so has the way that kids interact with them.

When engineers talk about this, they say machines have lost their *transparency*, their ability to be divined simply, easily, and visually. If you took apart your mom's washing machine, as Kelley did in the 1960s, you'd find a metal tub whose gyrations were accomplished by pulleys, belts, and gears. You fixed the thing by pulling it apart and inspecting each piece and replacing the worn ones. If electronics was your thing, you assembled your metal detector kit with dime- or quarter-sized components you could easily manipulate: You soldered capacitors and transistors to pegboards, and coiled copper wire around your palm and carefully bolted it into the chassis head. Today, if you take apart a CD player or a disk drive or a remote control toy truck, there are some moving parts, but the real brains of these things, computer chips, are too small to read or to repair manually. They're *opaque*. You know what they do; you just can't see how they do it.

Once kids started growing up with a houseful of opaque items, goes the theory, they lost interest in taking them apart. Today it's more fun to see what an intact machine can do. More fun jerking joysticks, pounding keyboards, taking virtual rides. The result? Two decades of kids have become proficient at running, loading, and hacking software to the detriment of their hands-on skills. Their ignorance has not gone unnoticed, even at MIT. "The students are a little more voyeuristic, living a little more virtually and not getting their fingernails dirty," says Michael Hawley, a professor at the Media Lab. "I do think that's true, but it's hard for me to measure and I can't back it up with numbers. But it's a sense that I do have."

3. Building Tomorrow's Castles

The quintessential course—the one that spawned countless worthy imitators and a crass TV show—was a competition begun at MIT in 1969. Back then, undergraduates taking 2.70, Intro to Design, were handed boxes of junk and challenged to design a machine over the course of a semester that would move continuously down a ramp in precisely fifteen minutes. Every team's solution was different, and generations of students have gathered in the school's gym to watch the battle between the rickety and sure-footed. The tradition continues to this day. In the '70s, the box contained cardboard, paper clips, rubber bands, a bag of sand, and a peanut (for lubricating parts). In 2002, the box contains sheet metal, motors, pistons, and joysticks, and students spend at least part of their semester using computer software to design the project.

That competition, which Flowers helped create as a grad student in the '60s, was exported to high schools in 1990 by Kamen, the Segway inventor. "I was looking for a way to create excitement in the culture about science and design," says Kamen. He found it. In 2002 alone, 20,000 teenagers from 650 high schools in the US and Canada stayed up nights and weekends building machines, in the hopes of snagging one of sixty-five college scholarships. High schools threw potluck supper fundraisers to send their ambassadors to the finals. Meanwhile, on cable TV, *Battlebots* features blood matches, in which designers wield machines like weapons, physically annihilating other inventors' machines. Winning designers get a toy deal from Hasbro. Kamen and Flowers deplore the show's violence.

Highbrow or low, robots are regarded as an excellent tool for luring hands-off kids into engineering. Elementary school kids seem to enjoy the idea of building a one-of-a-kind machine that holds meaning only for them. A robot that chops bananas. A robot that dips tortilla chips in salsa. A tug of war machine. A soccer-ball kicker. A volleyball trainer. An automatic writing machine. All of these were, in fact, built last spring by students at the Little Red School House in New York City using ingredients from the world of the mechanical: Legos, tape, sticks, felt, staples, and the electronic—programmable chips called Crickets.

Crickets sprang from the mind of Mitchell Resnick, an MIT Media professor who has been thinking hard about how to marry the digital and analog worlds in such a way that even young children can watch that union happening in front of them. "Since the technological revolution, kids can open a machine, but all you see inside is this black box," he say. "So we're trying to get kids to look at the boxes the way they really are: boxes that can perform work, but you program what they can do."

Resnick's first invention was the Programmable Brick, a 2-by-4-inch, easily handled chip that can be programmed repeatedly by kids using menu-driven software. Crickets, the second generation of that technology, are smaller chips that have a pair of motors and some bus ports for peripherals. A child writes software instructions on a computer screen, then beams the commands to the Cricket as easily as he or she could click on a TV remote. A Lego-built vehicle will behave like a car—go forward, left, right, and reverse—once a 1 1/4-by-2 1/4-inch Cricket is programmed, snapped into the body, and attached to its axles. Crickets and Legos let kids play in both worlds: the electronic and the mechanical. Legos are particularly good classroom tools because you can alter the parts of your machine by hand. In a perfect world, a child would later make the transition to woodworking or machine-shop tools.

Emma and Gloria, a pair of sixth graders in Mr. Smith-Welch's class at the Little Red School House, have built a toy to entertain—or more likely bedevil—Emma's dog. The robot is a little car that drags around a rubber ball. A little dude who resembles *Sesame Street*'s Bert sits up behind the dummy controls of the car in what looks like a lifeguard's beach chair. On a spring morning, Emma, in a hooded sweatshirt, and Gloria, in a soccer T-shirt, hair in braids, sit at their Mac designing a software program. All they have to do is tell the Cricket's left and right motors when to go on and off, and how many seconds to wait. They type the LOGO directives:

AB, ON BEEP BEEP
    Wait 12
    A, off
    Beep beep
    Wait 33
    Beep
    Ab, on
    Beep beep
    Wait 6
    Ab, off
    Beep beep beep beep

. . .

"WHY ALL THE BEEPS?" SOMEONE ASKS.

Emma shrugs.

"Just for fun," Gloria says.

And it is, for a while, but the machine just runs in a circle for a little bit before stopping. As the girls play with it in the hallway outside their class, Michael Smith-Welch stops by to watch. He's an energetic young man with a background in art and sculpture. "Why don't you attach a touch sensor to the ball?" he tells them. "That way, when the dog bites it, it'll stop."

The girls follow him over to a benchtop where hundreds of Lego parts lie waiting in a heap. Smith-Welch picks through the detritus until he finds an old robot, and starts detaching the part they'll need, a pair of wires attached to a metal clip. Depress the clip and it'll send a message to the Cricket. Weekends and nights when he's not grading papers, Smith-Welch builds these extra parts from stuff he picks up at Radio Shack. And he talks a lot about what he builds—there's not an eleven-year-old in the class who doesn't know a light sensor from a touch sensor.

"You know to rewrite the program so the Cricket stops, right?" Smith-Welch says.

The girls nod. They look nervous. Emma shifts her weight to one leg; her arm hugs her stomach. "What do we attach it to?" says Gloria.

Smith-Welch smiles like a mischievous elf. "Ah. It's a mechanics problem, isn't it?"

Not all schools are created equal. Smith-Welch makes all this work in his science lab because he's an exceptional teacher, because he's hooked up with the guys at MIT's Media Lab, and because Little Red's private school budget is large enough to allow him to buy enough Lego sets to approximate a junk pile. Not all schools can do that. Worse, not all schools have the imagination to let kids play. The second a teacher breaks out a bland, safe Science Adventure kit, plus worksheets, the jig is up. Play, managed by adults, quickly becomes work, and a kid will see right through the scheme.

So, for most kids, it's going to come down to what they do

on their own, with their pals, or with their parents. "If you have a kid," says Flowers, "why not go to the junkyard, pick up a lawn mower, take it apart and see how it's built? That's what I'd do. You take apart one thing a week and see what's behind the cover. Or before you throw something away, why not take it apart?"

Not long ago, Kelley, who, in addition to teaching at Stanford, runs IDEO, the world's largest and most innovative industrial design firm, embarked on a similar feat of exploration with the littlest engineer on his payroll: his five-year-old daughter. When she announced she wanted a playhouse shaped like a castle, the guy who designed the Palm V Pilot decided to treat the experience as he would a work project, progressing from research to prototype to finished design. They watched *Sleeping Beauty*. They researched castles on the web. They noted the color of castles, the bricks, the flags, the moats. The two of them built a little model from sticks, cardboard, and glue, then moved on to the real thing. "I had to drive most of the nails, but she was out there with a hammer and a paintbrush," he says. "The whole thing is to get these kids into projects that they love and make them hands-on. Make them do everything themselves, not just the thinking. Not just the planning and designing but the construction as well. They build confidence and they start to feel good."

Kelley's instinct to reach for everything from hand tools to keyboards and videos is on the mark. Sometimes the old farts sound like they want to ditch the computers. That would be a tragic mistake. Computers are here to stay, and it's important for young people to master them. "There are a lot of people who think we should go back," says Flowers. "I don't think they're thinking very carefully about the problem. I think we need to create a new mix. Going back is the wrong solution. Taking full advantage of the right mix of new things is more important. If we do that, I think we'll see kids who are comfortable using all kinds of tools. We'll see new a kind of nerd pride."

So tell the kids you know. You need to write code *and* mess with the hardware. You need to gain proficiency in both realms because the stakes are growing ever higher. What happens when there's an accident, when nine men need to be extracted from a mine? What happens when a city is brought to its knees, as one was in September 2001, and the computers are down and people look to you because, hey, you're supposedly good with your hands?

God forbid that ever happens again. But maybe someday you'll want to work for an influential design firm such as IDEO in Silicon Valley. You'll walk into a room to meet the company president, a proud nerd who teaches college courses on the side and grew up taking things apart. You want a job from this man, and you'll have to answer truthfully: Did you tinker as a kid? "The ones that say yes, I hire," says Kelley. "The ones that don't, I don't."

If you took things apart as a kid, you'll thank your lucky stars. And when you gain entrance to this playhouse, you'll sit at a table during brainstormers, and someone will bring out The Box. It's filled with foam, little mechanisms, parts of toys, retractable razors, miniature fans, and you and your coworkers will fiddle with these things as you talk, looking for inspiration. Your college buddies are off worshipping at the Church of the Opaque, but you and your new friends are pulling things out of the Box. Do you know what this is? someone says. Yeah, it's a convex lens from a telescope. And this? An old transistor. And this? A motor from an old erector set, you say as you reach for it. See the wires? You haven't seen one in years, but your fingers close over it, and it fits in your hand like an old friend.

BOOK AND MEDIA REVIEWS

All the reviews that follow first appeared in national science magazines in the US.

G*hosts of the Abyss*, directed by James Cameron.

James Cameron is a man obsessed. His blockbuster film *Titanic* won eleven Oscars in 1998, but the director clearly can't get enough of the doomed ocean liner. Three years later, he returned to the hulking ghost ship of the North Atlantic, accompanied by historians, scientists, and the Russian crew of the Akademik Mstislav Keldysh, the world's largest oceanographic research vessel. The result is *Ghosts of the Abyss*, an IMAX 3-D documentary that provides an unprecedented peek at the world's most famous grave marker. This time, there are no panpipes, no sophomoric love stories, and, thank goodness, no Celine Dion.

Technology is the crowning glory of the film. Cameron spent bags of lucre from the box-office triumph of *Titanic* to design cameras that could withstand crushing sea pressure and navigate the wreck in ways human divers could not. The onscreen heroes are a pair of swimming robots named Jake and Elwood, who come equipped with lights, fanlike propellers, cameras, and all the spunky charisma of R2-D2. Small enough to squeeze into holes in the ship's hull, the bots nimbly navigate the dead ship's unpredictable hazards and beam back images to the team's three-man submarines. In this way we're privileged to behold parts of the luxury liner that have not been seen since it sank. The artifacts include leaded glass windows in the main dining room; the brass bed where the unsinkable Molly Brown slept; the Marconi wireless room that sent the ship's last anguished cry for help; and the great coal engines stilled by ninety-one years of brine.

Obscured by barnacles and sea organisms, shipwrecks can make incoherent footage. To aid comprehension, Cameron resorts to split screens, computer animation, and ghostly actors who re-create the ship's final moments. All this interpretive icing threatens to sink the ship a second time. The most powerful stuff is right in front of us, if only the filmmakers would shush and let us watch the show. One realizes that it was not really hubris that sent this great work of human hands to its death but Mother Nature.N ow she is finishing the job. An onboard microbiologist waxes poetic about the millions of sea dwellers—crustaceans, seaweed, and primitive fauna—that are slowly digesting the ship and returning it to the earth.

It's a powerful message. The beauty of *Ghosts of the Abyss* is that it immerses you in larger-than-life underwater footage you won't easily forget. You slip on IMAX 3-D glasses and pay an hour-long visit to an inaccessible realm, where a tragic vessel speaks more clearly than its explorers ever could.

~

SEXY? GLAMOROUS? COOL? IT MUST BE FORENSIC science.

A medical examiner dips a ladle into a tall stainless-steel stockpot and casually lifts a severed human head from the bubbling liquid. To determine the cause of death, he must wait a little longer, until all the residual flesh drops off. So he plops the ladle back into the stew. No muss. No fuss. Just business as usual on the Emmy Award–winning show *CSI: Crime Scene Investigation*.

These days, it's impossible to channel-surf without stumbling over a corpse. Nearly every night, on such shows as *CSI: Miami, NCIS, Crossing Jordan*, and *Bones*, we are treated to glimpses of the human form in various stages of decay. Under the dreamy blue lights of these fictional labs, the dead are eviscerated in graphic detail—bodies splayed wide, sternums snipped with garden shears, skulls popped to disclose brain matter, faces and chests peeled back like chicken skin.

Ostensibly, these programs attract huge audiences because they reveal forensic science to be a fascinating field. In fact, the stylized gore, throbbing pop scores, double entendres, and hot pastel colors are giveaways that something far craftier is going on: a subtle game of titillation. When *CSI: Miami*'s Dr. Alexx Woods (played by the tall, striking actress Khandi Alexander) gently strokes a corpse, speaking to it in her hushed, comforting tone, or when the cast of *CSI: New York* regale viewers with clinical talk of semen, saliva, and the swabbing of intimate orifices, even the daftest viewer must concede that we are being treated to a new kind of erotic striptease. Death has become the new sex.

These are make-believe shows about a dream world where death looks scary but really isn't. Viewers are assured that even if their physical selves are destroyed in a hurricane, tsunami, earthquake, mudslide, or terrorist attack—the all-too-real horrors of the modern day—they will remain gorgeous, even in death. A forensic scientist will caress your corpse and turn your passing into a crossword puzzle, something to be decoded, and maybe even kind of cool.

The Whale and the Supercomputer: On the Northern Front of Climate Change, by Charles Wohlforth.

A team of scientists studying global warming in Alaska hire an Eskimo to lead them into the Arctic wilderness. Each day during breakfast, their guide leaves the main tent and returns with a weather forecast, ostensibly after studying the skies. Day after day, his predictions prove unerringly accurate. Finally, one of the scientists asks to accompany the Eskimo in order to learn his arcane method for divining the weather. "Sure," he replies. "I'm just going over to my tent to listen to the seven o'clock forecast."

The story may be apocryphal, but to Charles Wohlforth it carries a ring of truth: For the lowdown on climate in the Arctic, he argues, it often makes more sense to consult the plainspoken wisdom of the indigenous Alaskans rather than rely exclusively on abstruse data gathered by well-meaning but often clueless scientists. After all, the Arctic natives have spent the past few decades watching the climate warm before their eyes. Since the 1950s, the average winter temperature of Alaska's interior has risen by seven degrees Fahrenheit. Precipitation increased by almost a third between 1968 and 1990, and entire villages have been relocated due to rising water levels. Meanwhile, as homes sink into the muddy, melting earth, insects such as the spruce bark beetle, whose numbers were once kept in check by cold weather, have been decimating forests. Staying alive as this once-frozen world turns to slush means keeping your wits about you and gleaning information about the environment from every available source, including the radio, the retreating ice, or your whaler uncle, Joe.

Wohlforth, a former reporter for the *Anchorage Daily News* who followed climatologists across Alaska, doesn't dismiss the scientific data they have meticulously collected. But he does suggest that the scientists' ability to understand climate change—whose impact is far more pronounced in the Arctic—is hampered by their tendency to sneer at anecdotal evidence. Huddled close to their $40-million-dollar supercomputers, climatologists can't even correctly model Arctic snow cover, even though understanding the white stuff is critical: The Arctic's blanket of whiteness reflects 80 percent of the sun's warmth. Lose snow and we absorb sunlight, at our peril. As for this research team's Holy Grail—predicting the change in average global temperature—it begins to look more and more like an unreachable, even meaningless, goal. As Gerard Roe, a modeler at the University of Washington, confides, "At some point someone is going to ask, 'What have you got to show for forty years of research?' And I'm afraid the answer will be, 'Not that much.'"

Meanwhile, the native Iñupiat congregate in community centers and homes, swapping stories over pots of boiling walrus meat and functioning like a living, breathing supercomputer. Several times, we learn, indigenous knowledge has surpassed that of the scientists: In 1977 the International Whaling Commission ordered whaling stopped, saying bowhead whale numbers were alarmingly low. The Iñupiat insisted that the figures were wrong. In the 1990s, oil industry spokesmen claimed that their deep-sea soundings would not unduly disturb whales. The Iñupiat dissented. Both times, scientists redesigned their studies to incorporate indigenous knowledge; both times they found that native hunches had been correct.

Despite this, the scientific establishment still finds it hard to blend native understanding into its work. In one scene, a climatologist seeks to make the Iñupiat partners in his research, but his superiors deem the plan "bad science." Shame on them. If the geniuses hope to get us out of this mess, they must learn to embrace indigenous insight. As one researcher admits, if you're heading for thin ice, you want a native elder, not an ice scientist, at your side.

~

Postcards From the Brain Museum: The Improbable Search for Meaning in the Matter of Famous Minds, by Brian Burrell.

Lord Byron, the poet who first nudged Mary Shelley to pen the novel *Frankenstein*, inadvertently inspired another myth: that the root of greatness lay in an oversize brain. To promulgate this notion, Byron did nothing more than drop dead. Following Byron's demise in 1824 in Greece, a curious physician removed the poet's gray matter and discovered it weighed "six medicinal pounds," roughly 25 percent heavier than the average brain. News soon spread through the medical community, spawning a cultlike investigation into the connection between anatomy and intelligence.

Postcards From the Brain Museum is a tragically funny history of this flawed paradigm. It tells the story of the golden age of brain collecting, from the late nineteenth century to the mid-twentieth, when famous people bequeathed their brains to science, expecting posthumous confirmation of their brilliance. Such study of "elite brains" consisted largely of inspecting physical attributes such as size, surface fissures, or cellular content, each of which was considered, at one time or another, to be the crucial indicator of intelligence. Great men had teeny brains too, but zealous researchers could always explain away that inconvenient fact by claiming that the number of these was insignificant.

Brian Burrell, a mathematician at the University of Massachusetts in Amherst, travels the world to inspect these shriveled relics, which languish in the obscure back rooms of museums and universities. The charm of his book lies in its revelations about famous figures touched by this futile endeavor. Walt Whitman became an adherent of the goofy phrenology movement in the United States, only to have his brain destroyed, probably when it was dropped by a clumsy handler. Einstein's was harvested along with his eyes, now rumored to be in a New York City bank vault. Lenin's was sliced wafer thin, mounted on slides, and scrutinized to support the legitimacy of Communism. France's Mutual Autopsy Society, whose members pledged to leave each other their brains, behaved like the Ouroboros, a mythical self-consuming snake: Young members harvested the old, only to be harvested themselves decades later. And Cesare Lombroso, the father of criminal anthropology, concocted the wrongheaded idea of a criminal mind, which coincidentally influenced the plot of every Frankenstein film ever made.

Nowadays, of course, brain science follows more fruitful avenues such as the chemistry of neurotransmitters and the analysis of brain function using CT scans and magnetic resonance imaging. Pickled brains—elite, criminal, or not—have been consigned to dusty closets, because, as Burrell says, researchers finally accepted a prosaic truth: All brains look alike.

Spook: Science Tackles the Afterlife, by Mary Roach.

Does the human soul exist, and if so, how much does it weigh? In April 1901, a surgeon named Duncan Macdougall devised a morbid experiment at the Consumptives' Home in Dorchester, Massachusetts, in order to find out. He placed a dying tuberculosis patient on a platform scale rigged with a cot. At the moment of death, Macdougall hypothesized, the scale should register a drop in weight when the ghost shucked off its earthly body. Their first patient in place, Macdougall and his team took up their grim posts. "In the place of the more usual bedside attitudes of grief and pity, the men assumed an air of breathless, intent expectancy," writes Mary Roach. "I imagine you see this on the faces of NASA engineers during countdown and, possibly, vultures." The first patient's soul clocked in, er, out at the now-legendary 21 grams. Sadly, Macdougall never managed to duplicate that data on later patients.

As Roach explains in this hilarious book—a follow-up to her equally funny *Stiff: The Curious Lives of Human Cadavers*—nailing down the attributes of the human soul has been a legitimate pursuit for scientists since the days of Aristotle. As recently as the 1920s, journals such as the *Lancet* and *Scientific American* routinely published papers on now dubious topics such as telekinesis, mediumship, and of course, soul-weighing.

Take the case of W. J. Crawford, a professor of mechanical engineering at Queen's University in Belfast, who in the 1920s cranked out five books on séances. After logging hours investigating the phenomenon, Crawford insisted that "the spirits" spewed ectoplasm—ghostly residue—from the bodies of mediums. Roach inspects the photos in his books and wonders how Crawford could have been duped for so long: these mediums were clearly gifted regurgitators, disgorging cheesecloth, silk, or even sheep's lungs from their mouths or other orifices.

Modern scientists seem to possess a stronger sense of skepticism. At the University of Virginia, researchers have installed a laptop computer atop a tall bank of cardiac monitors in a hospital operating room in which patients undergoing defibrillator implantation routinely have their hearts stopped. No living person in the operating room can see the twelve simple images (a frog, a plane, or a doll, among others) flashing on the laptop's screen, since it faces the ceiling. But if you had died, and your soul had floated to the ceiling—a scenario described by legions of near-death survivors—you would have no problem later describing for researchers the images on the laptop. So far in this study, no one has.

At Laurentian University in Sudbury, Ontario, neuroscientist Michael Persinger has induced paranoia in humans by subjecting them to low-level pulses of electromagnetic fields. The creepiness one feels in an old house, he posits, is more a matter of ungrounded wiring and overloaded circuit boxes than spooks. Unlike the book's historical figures, who come off as cranks, modern investigators appear as credible researchers with unconventional side interests. Perhaps only in death does one's paranormal work seem kooky. There's a lesson in that for scientists: don't die. Or, if you do, return often to defend your theories.

~

Empires of Light: Edison, Tesla, Westinghouse and the Race to Electrify the World, by Jill Jonnes.

Thomas Edison, the quintessential American tinkerer, is often held up as a role model for schoolchildren. But the books that celebrate his inventions, which include the first lightbulb and the first phonograph, don't mention his unusual interest in the art of electrocuting dogs. In 1888, Edison and his cronies offered local children 25 cents for every mangy pooch they could drag into his labs in West Orange, New Jersey. One by one, the mutts were caged, wired, and zapped—first with direct current, an Edison invention, and then with alternating current, the darling of George Westinghouse. Because it took 1,000 volts of DC to fry the ill-fated dogs but only 300 volts of AC, this was proof—to Edison, at least—that AC was more hazardous. This was delicious news to Edison, since he was at the time locked in a battle with Westinghouse to wire the nation (and potentially the world) with electricity. At stake was the future of American industry: Each titan stood to gain millions of dollars, but only if his system proved the best.

Historian Jill Jonnes re-creates this venomous rivalry in a delightful book that may remind readers of E. L. Doctorow's novel *Ragtime*, in which real-life people pop up in fictional settings. But *Empires of Light* is no fiction; it's a meticulously researched narrative in which famous people go baying after an elusive goal: to power cities by harnessing a hidden force wrested from the atmosphere. Such a quest was once viewed as the work of the gods. But Jonnes's subjects—Edison, Westinghouse and the eccentric Serbian-American genius Nicola Tesla—are all too mortal: brilliant men driven by passions as mercurial as the invisible ether they pursued.

Edison's macabre tableau of dead dogs—a parade that led to the creation of the nation's first AC-powered electric chair—was part of his plan to brand alternating current, in which the flow of electrons periodically reverses direction, as too dangerous. He favored a system that supplied power directly from a central station to consumers close by. Though safer, it lacked efficiency: sometimes power ebbed, and the cost of erecting so many stations and copper-heavy grids made DC practical only for densely populated cities. Westinghouse, by contrast, had the technology to snake wires from remote power plants to cities hundreds of miles away, via transformers that could "step down" the high-voltage—and potentially lethal—AC power leaving the plant to a level safe for homes and businesses. The AC grids were also more efficient: one plant instead of a dozen meant that they required far less copper.

Although Edison plied his folksy wit to charm reporters and woo investors, it was Westinghouse's scientifically and economically superior platform that ultimately won out, in part due to the ingenuity of Nicola Tesla. Early electrical designs often faltered in their power output when poorly machined whirling parts scraped against each other. Tesla's solution was to position two separate magnetic fields on the same motor; when one waned, the other surged, producing a single stream of power from two alternating waves of current. Westinghouse licensed the technology and took AC to the masses. Meanwhile Tesla was relegated to oblivion. Fastidious, foppish, and phobic, he neglected his patents, sank into poverty, and spent much of his later life pursuing crackpot schemes, even claiming once that he had intercepted messages from Martians.

Jonnes has a remarkable knack for re-creating the sounds and smells of the nineteenth century's cluttered labs and opulent boardrooms. There are no massive university budgets here; just fat men with stogies bankrolling one dream after another. In the end, this world is a little like watching a primitive version of a cable TV finance show: money guys howling for profits; journalists fanning rumors that the science is going nowhere; and the geniuses at the center chasing a quarry only their eyes can see.

Greek Fire, Poison Arrows & Scorpion Bombs: Biological and Chemical Warfare in the Ancient World, by Adrienne Mayor.

In the run-up last spring to the US invasion of Iraq, a parade of pundits appeared nightly on the news, spouting ideas honed by a century's worth of war atrocities. Good warriors fight fairly, they insisted. Chemical weapons are the coward's way out. Our side *does* research biological weapons—but only for defensive purposes. Though these notions may seem novel, they are, in fact, ancient history. The first-century AD Roman historian Florus, for example, lambasted a general for poisoning wells, while his contemporary Tacitus praised the Goths, who spurned poison and chose pitch-dark nights for attack.

Thus, if there is any lesson to be learned from *Greek Fire,*

Poison Arrows & Scorpion Bombs, it's that there's nothing new about biological and chemical weapons—or in the universal hypocrisy surrounding their use. During the Vietnam War, American forces sprayed civilians with napalm, a jellied gasoline that burned at more than 5,000 degrees Fahrenheit, consuming clothing, skin, and bone. It was widely assumed that Harvard scientists had invented the stuff in 1942. Few people realized that liquid incendiaries—naphtha, pitch, or quicklime—were used as far back as 875 BC. Later, in the seventh century AD, a Syrian engineer named Callinicus invented a terrible cannon that pumped a fiery naphtha mix through bronze tubes at ships in battle. Dubbed Greek fire, it doomed its target instantly.

Mayor, a classical folklorist, recounts in lively, sometimes darkly comic detail the diabolical stratagems devised by devious warriors for tactical ends: arrows dipped in toxins; jewel-encrusted urns booby-trapped with plague-laden garments; and a host of dirty tricks involving snakes, stinging beetles, and venomous frogs. Ironically, the more humans resorted to these insidious tools of destruction, the more the concept of war became idealized. During the brutal Peloponnesian War, the Athenian historian Thucydides condemned atrocities against noncombatants and praised "courage and sheer strength" over "scientific methods." Yet the chicanery not only prevailed but bred arms races that continue to this day. Defeated once by war elephants in the second century BC, the Macedonian prince Perseus ordered his men to build wooden pachyderms from which musicians blew trumpets. In this way, warhorses became inured to the sight and sounds of the tusked raiders. In other battles, besiegers unleashed pigs and camels to frighten elephants and horses, whereupon the generals exposed their stables to these animals and their offensive odors. By AD 1398, the Turkic conqueror Tamerlane had hit upon a new trick in his onslaught on Delhi: releasing flaming camels across the battlefield.

Such insidious tactics could backfire, though. As Mayor points out, biochemical weapons are the ultimate double-edged sword. Hercules buried Hydra's head because he knew it was immortal. Today we talk of stockpiling our nuclear and chemical weapons in desert repositories. We can hide them, but we can never rid them completely of their power to kill. This is the hard lesson of history that we moderns have yet to learn: If you unleash a flaming camel, sooner or later it'll go berserk and run straight back into your camp.

High Tech Trash: Digital Devices, Hidden Toxics, and Human Health, by Elizabeth Grossman.

If you sleep comfortably having lugged your old computer to a hazardous waste collection site, certain it is being safely picked apart and recycled into next season's mp3 player, wake up. Chances are, your trash was shipped to a rural province in China, where poor, sandal-clad women melt circuit boards in skillets to harvest the precious metals. These workers wear no protection and breathe fire retardants, dioxins, and furans.

As Elizabeth Grossman reveals in her dense and damning book about the afterlife of techno-trash, much of what we cast away is improperly disposed of—regardless of our efforts—and ends up leaching toxins into the air, water, and soil. Between 2003 and 2010, as many as three billion units of consumer electronics generated by the States will head to the global junk pile. If current trends persist, only about 10 percent of this will ever be recycled.

Part of the problem is that electronics require costly dismantling and sorting of their ingredients. In Europe and Japan, recycling is mandatory, forcing manufacturers to shoulder the financial burden of recycling waste and compelling them to design less-toxic products. Not so in the United States. Lacking tough tech-recycling laws on the federal level, local governments fashion regulations that are hard to follow. As a result, e-waste is discarded along with junk mail and food scraps and ends up in landfills where shattered screens spew lead and mercury and circuit boards leak cadmium, a probable carcinogen.

"Microchip circuitry may be as invisible as the network of nerves on a dragonfly's wing," Grossman writes. "And whole libraries may appear on our desktop screens apparently out of thin air, but unless some radical changes are made in the way we design and produce our information age gadgetry, its ecological footprint will never really be reduced."

Body Brokers: Inside America's Underground Trade in Human Remains, by Annie Cheney.

Every human life is precious, and so, it seems, is every human body. A cadaver fetches up to $5,000 if offered to the right buyer, but like a hijacked Corvette, the same corpse will earn far more if it is sliced up and its parts sold separately: $600 for the brain, $900 for the now-empty head; $3,000 for a headless, armless, and legless torso; $850 for an elbow; $280 to $500 each for miscellaneous organs.

Cadavers have long played a critical role in research—to teach anatomy in medical schools, for example—but donors who leave their bodies to science may be disconcerted to learn that their organs are often diverted to more entrepreneurial ends. As Annie Cheney reveals in *Body Brokers: Inside America's Underground Trade in Human Remains,* today's markets include training seminars for physicians seeking to study new surgical techniques, technicians designing the next generation of "minimally invasive" procedures, and tissue-engineering firms that repackage body parts—skin, ligaments, heart valves, and bone—for transplants.

What donors and their families don't realize is that medical schools and universities typically sell remains to the highest bidder when their inventory overflows. Body parts are also callously stolen from hospital mortuaries and funeral homes, as in the case of veteran broadcaster Alistair Cooke, who died of cancer in 2004 at the age of ninety-five. Police say Cooke's remains were plundered on their way to a crematorium in New York City by a gang of modern-day grave robbers who sold the bones on the transplant market, using fake documents that falsified the "donor's" age and cause of death.

Cheney argues that these crimes have become more common because Americans are increasingly distanced from death. When a family member dies, he or she is immediately consigned to the care of strangers—and left vulnerable to the unscrupulous. This horrifying trade demands closer scrutiny, not only by federal and local authorities, but also by anyone who sends a loved one down the lonely path toward the grave.

A REQUEST FROM THE AUTHOR

Before you go, permit me to mention two small things...

1. If you enjoy my books, I'd be extremely grateful if you'd leave a brief review of any of the ones that you have read at your retailer's site. Reviews are one of the best ways a reader can help an author. Reviews help other readers find out about the book.

2. My newsletter is the best way for me to stay in touch with readers. Please consider joining. You'll always be among the first to learn about upcoming books and deals. As a welcome gift, you're entitled to download any or *all* of the ebooks in my Starter Library, completely free. Your contact info will never be sold or shared with anyone. You can easily opt-out at any time. Did I mention it's free?

VISIT JOSEPHDAGNESE.COM TO SIGN UP

ACKNOWLEDGMENTS

All articles copyright by Joseph D'Agnese. All rights reserved. "The Scientist and the Sociopath" © 2004 first appeared under slightly different form in *Wired*. Portions of this article appear with the kind permission of Gordon Rugg. "Last of the Big Bang Men" © 1999 first appeared under slightly different form in *Discover*. "Where Chimps Go to Die" © 2002 first appeared under slightly different form in *Discover* and *Best American Science Writing 2003.* "The Medical Detective" © 2007 first appeared under slightly different form in *Discover*. "The Organ Makers" © 2001 first appeared under slightly different form in *Discover* and in *Best American Science Writing 2002.* "Organs in the Shop" © 2005 first appeared under slightly different form in *Discover*. "The Lady in the Suitcase" © 2002 first appeared under slightly different form in *Discover* and in *Reflections on Anthropology: A Four-Field Reader, 2003.* "The Scientist Who Threw the Light Switch" © 2003 first appeared under slightly different form in *Seed*. "The Weather Inside You" © 2000 first appeared under slightly different form in *Discover*. "Patriots Aboard the Big Blue Bus" © 2010 first appeared in slightly different form, first in *On Earth*, then in *The Utne Reader*. "The Town Changed by Wind" © 2007 first appeared under slightly different form in *On Earth*. "Mars Bound" © 2003 first appeared under slightly different form in Discover. "Do More With Less" © 2002 first appeared under slightly different form in *Discover*. "Know Thyself" © 2000 first appeared under slightly different form in *Discover*. "Worshipping at the Church of the Opaque" © 2002 appeared in slightly different form in the September 2006 issue of *Child*. All reviews © 2001-2011 first appeared under slightly different form in *Discover*.

ABOUT THE AUTHOR

Joseph D'Agnese is a journalist and author who has written for children and adults alike. He's been published in the *New York Times,* the *Wall Street Journal, Wired, Discover,* and other national publications.

In a career spanning more than twenty years, his work has been honored with awards in three vastly different areas—science journalism, children's literature, and mystery fiction.

His science articles have twice appeared in the anthology *Best American Science Writing.*

His children's book, *Blockhead: The Life of Fibonacci*, was an honoree for the Mathical Book Prize—the first-ever prize for math-themed children's books.

One of his stories won the 2015 Derringer Award for short mystery fiction. Another of his short stories was selected by mega-bestselling author James Patterson for inclusion in the prestigious annual anthology, *Best American Mystery Stories 2015.*

D'Agnese's fiction has appeared in *Ellery Queen's Mystery Magazine* and *Alfred Hitchcock's Mystery Magazine.*

D'Agnese lives in North Carolina with his wife, the *New York Times* bestselling author Denise Kiernan (*The Girls of Atomic City*).

amazon.com/author/josephdagnese

bookbub.com/authors/joseph-d-agnese

goodreads.com/JosephDAgnese

Made in United States
North Haven, CT
04 November 2026

21767249R00109